ADVANCED
FREQUENCY SYNTHESIS
BY PHASE LOCK

ADVANCED FREQUENCY SYNTHESIS BY PHASE LOCK

WILLIAM F. EGAN
Santa Clara University

A JOHN WILEY & SONS, INC., PUBLICATION

Library of Congress Cataloging-in-Publication Data:

Egan, William F., 1936- author.
  Advanced Frequency Synthesis by Phase Lock / William F. Egan.
    p. cm
  ISBN 978-0-470-91566-0 (hardback)
 1. Frequency synthesizers. 2. Phase-locked loops. I. Title.
  TK7872.F73E298 2011
  621.3815'486–dc22

                                        2010049579

Printed in Singapore

oBook ISBN: 9781118007716
ePDF ISBN: 9781118007693
ePub ISBN: 9781118007709

10 9 8 7 6 5 4 3 2 1

To
Kimberly
Melody
Li-Chuan

CONTENTS

3 OTHER SPURIOUS REDUCTION TECHNIQUES   39

4 DEFECTS IN $\Sigma\Delta$ SYNTHESIZERS   55

PREFACE

This book is a continuation of *Frequency Synthesis by Phase Lock*, 2nd edition (*FS2*), for the purpose of expanding coverage of techniques that have become more important since its publication and to introduce new techniques. The area of increased importance is primarily sigma–delta ($\Sigma\Delta$) synthesis, while all-digital synthesizers represent the emerging technology, and diophantine synthesis shares some of both characteristics. We will depend on *FS2* to define the assumed technical background, without requiring that it be acquired there. We will also make extensive use of Simulink® models, without requiring their use by the reader.

Frequency synthesizers that employ $\Sigma\Delta$ modulators offer the potential to provide wide bandwidth simultaneously with fine resolution (step size), plus other advantages. When *Frequency Synthesis by Phase Lock*, 2nd edition, was published in 2000, they were rare and considered exotic. Since then, they have become widely used in high-production consumer electronics, where they are valued for their compatibility with IC technology. The theory of their operation was discussed in *FS2* but, as would be expected, much valuable information has become available since that time. This, and their new prominence, now justifies greater attention.

We will consider the fractional-N synthesizer, of which $\Sigma\Delta$ is a special case, as a whole, moving from simple fractional-N, through analog compensation, into $\Sigma\Delta$ modulation. We will do so in a measured fashion, as if by experimental observation, using Simulink models as the experimental objects. This will permit flexibility in observing the effects of variations in design and parameters. Moreover, these models will sometimes provide what appear to be discoveries beyond what is generally known.

After a brief introduction in Chapter 1, we will move to fractional-N and basic $\Sigma\Delta$ synthesis in Chapter 2. Here, we will see how spurs are produced at the synthesizer output by the basic process and learn some means for overcoming them. We will study additional methods for spur suppression in Chapter 3. Then, in Chapter 4, we will investigate how various defects contribute to spurs and other imperfections in the synthesized signal. Here, we will analyze synthesizer configurations and see how to trade off the various noise sources by choosing loop parameters. These synthesizers will all use the popular MASH configuration; we will have a look at other architectures in Chapter 5. Chapter 6 will describe the Simulink models that provided data for the preceding discussions and which can be downloaded from the web, much as MATLAB® scripts were made available in *FS2* (there are also new MATLAB scripts), and provide guidance to the reader who wishes to make use of Simulink models for further exploration. With Simulink models, we can answer questions that have not yet been asked.

Chapter 7 deals with another method for achieving wide loop bandwidth simultaneously with fine resolution, one that has been more fully developed since *FS2* and that is called the diophantine synthesizer. Chapter 8 further describes the analysis of synthesizers that rather than using these more sophisticated methods, push their bandwidths close to the (sampling frequency related) limit.

Chapter 9 introduces a newly important area, perhaps one that is currently "considered exotic," but is next to "become widely used in high-production consumer electronics," where it is valued for its compatibility with IC technology, all-digital frequency synthesis.

Much of the material is contained in Appendices, providing more in-depth discussion in a way that does not detract from the main flow and also permitting subjects to be referenced from multiple places in the main text. Many excellent sources are referenced throughout this work. Some are suitable for learning about these advanced techniques and some are better as references once the principles are understood.[*] Many related resources are available for download from the Wiley web site (see Appendix W). These include Simulink models, MATLAB scripts, spreadsheets, and executable programs, as well as the inevitable errata.

My intention is to provide a good learning source, taking the reader through an instructional journey while also providing material to permit expansion on what is learned. My hope is that the reader will also find this journey enjoyable.

ACKNOWLEDGMENTS

The creation of this work involves the help of many people and, I think, some rather fortuitous events. In June 2005, I attended a MATLAB seminar, in which Dick Benson demonstrated the simulation of $\Sigma\Delta$ synthesizers using the Simulink program and

---

[*]Among the fairly recent books that discuss $\Sigma\Delta$ synthesis, Crawford (2008) is most inclusive (including good historical background). Lacaita et. al. (2007), De Muer and Steyaert (2003), and Shu and Sánchez-Sinencio (2005) provide good information, including discussions of implementation. Among the papers referenced, Miller and Conley (1991) provides a fundamental introduction to $\Sigma\Delta$ synthesis; Hosseini and Kennedy (2007) provide a well-written introduction to $\Sigma\Delta$ MASH modulators in the Introduction of their paper. Staszewski and Balsara (2006) provide a book-length treatment of all-digital synthesizers.

models that were available at MATLAB Central. I was fascinated by the demonstration and requested, and was granted, access to the Simulink program through the MathWorks Book Program.

I began a series of simulations in an effort to better understand $\Sigma\Delta$ synthesizers. I experimented with the models, off and on for several years, a process that involved learning the intricacies of simulations and of DSP as well as the $\Sigma\Delta$ process. During that time, I benefited from the availability of the experts at The MathWorks and of technical literature through IEEE Xplore. In March 2010, Simone Taylor, who was taking over responsibility for my books at Wiley, contacted me to introduce herself. This eventually led to a review of the proposed book. The reviewers' comments were encouraging to Wiley and to me, so we proceeded with the project.

In June 2010, the Santa Clara Valley chapter of the IEEE Circuits and Systems Society announced a short course on state-of-the-art frequency synthesizers for CMOS technology. It featured lectures by three prominent authors (Liming Xiu, Paul Sotiriadis, and R.B. Staszewski; see References). This led me to study their works, as well as related works, and gave me an opportunity to hear and interact with them first hand.

I am grateful to many people, not only those whom I have specifically mentioned but also to those whose contributions were implied and also to those many authors whose works have been referenced in this book.

William F. Egan

November, 2010
Cupertino, California

SYMBOLS LIST AND GLOSSARY

The following is a list of terms and symbols used throughout the book. Special meanings that have been assigned to the symbols are given, although the same symbols occasionally may have other meanings, which should be apparent from the context of their usage. The meanings of other terms can be found through the index.

$|F|$	Magnitude of F.
$\angle F$	Phase of F.
ADC	Analog-to-digital converter.
ADPLL	All-digital phase-locked loop.
B_n	Noise bandwidth.
BW	Specified bandwidth of a spectrum analyzer.
c	Cycle(s).
CP	Charge pump.
CP&I	Charge pump and integrator (a charge pump feeding a capacitor).
crs	Noise bandwidth multiplier for all spectrum analyzers.
DA, DAC	Digital-to-analog converter.
dB_	Decibels relative to _.
dBc	Decibels relative to carrier (total power).
dBr	Decibels relative to 1 rad^2.
dB/Hz	Relative power density per hertz bandwidth (dBr/Hz is decibels relative to 1 rad^2/Hz, etc.).
DCO	Digitally controlled oscillator.
DFT	Digital Fourier transform.
DSM	Delta–sigma modulator (also DDSM, digital DSM, in this work).

EFM	Error feedback modulator.
EFMn	nth-order EFM.
ETF	Error transfer function, the ratio of the error (quantization noise) at the output to its value at its source.
f_{div}	Frequency at the frequency divider output, also called f_s in *FS2*.
f_e	Error frequency, $f_{ref} - f_{out}$.
fi	VCO initial frequency in hertz.
f_L	Unity open-loop gain bandwidth.
F_{LF}	Loop filter transfer function, du_2/du_1 ($K_{LF}F(s)$ in Fig. 1.1).
f_m	Modulation frequency.
f_{osc}	Oscillator frequency.
F_{out}, f_{out}	Output frequency from the synthesizer loop. Usually F_{out} is a fixed value, while f_{out} is a state variable indicating variations from initial or average value.
f_p	Pole frequency.
FPSD	Frequency power spectral density.
F_{ref}, f_{ref}	Reference frequency into the synthesizer loop. Usually F_{ref} is a fixed value, while f_{ref} is a state variable indicating variations from initial or average value.
F'_{ref}, f'_{ref}	Basic reference from which F_{ref} or f_{ref} is derived
fref	f_{ref} in hertz.
f_s	Sampling frequency. Used also for f_{div} in *FS2*.
FS2	*Frequency Synthesis by Phase Lock*, 2nd edition, see References.
f_z	Frequency of the zero in the loop filter.
G	Open-loop transfer function.
H	$G/(1+G)$.
I_{cp}, I_p	PFD pulse current.
K_{LF}	Loop filter gain constant, its DC gain (if finite).
K_p	Phase detector gain constant, $du_1/d\varphi_e$ (see Fig. 1.1).
K_v	VCO gain constant, df_{out}/du_2 (see Fig. 1.1).
$L(\Delta f)$	Single-sideband relative power spectral density at Δf from spectral center.
$\mathcal{L}(\Delta f)$	$S_\varphi(f_m = \Delta f)/2$. Equals $L(\Delta f)$ for small modulation index.
L_a	Length of (number of possible states in) an accumulator's register, 2^a.
L_{buf}	Length of a buffer, $2^{n_{buf}}$.
L_{segm}	Length of a waveform segment in samples.
L_{sequ}	Length of a repeated sequence in samples.
LF	Loop filter.
m	Peak phase deviation, called "modulation index."
MASH	<u>M</u>ultist<u>A</u>ge noise <u>SH</u>aping circuit. MASH-*abc* indicates that the stages have orders *a*, *b*, and *c*. That is, they have *a*, *b*, and *c* poles, respectively, in their z-domain transfer functions.
Mf	n_{fract}.
mux	Multiplexer.

N	Divider ratio.
$\overline{N}$	Average divider ratio.
N_a	Accumulator capacity $= 2^{n_a}$.
n_{buf}	Number of bits in a buffer.
n_c	Number of carries.
n_{fract}	Fractional part of $\overline{N} = N_{\text{int}} + n_{\text{fract}}$.
N_{fract}	n_{fract} times the capacity of the accumulator, the unnormalized input.
N_{int}, Nint	Integer part of $\overline{N} = N_{\text{int}} + n_{\text{fract}}$.
p	Order of $\Sigma\Delta$ modulator (P in *FS2*).
PD	Phase detector.
PFD	Phase frequency detector, commonly including the charge pump.
Phase margin	The added phase lag that will cause $\angle G(f_x)$ to equal $-180°$ when $|G(f_x)| = 1$; thus, $[180° + \angle G(f_x)]$. May also refer to the potential phase margin, $[180° + \angle G(f)]$.
PLB	*Phase-Lock Basics*, see References.
PLL	Phase-locked loop.
PM	Phase modulation, phase margin.
PSD	Power spectral density.
PSD_m	m-sided PSD ($m = 1$ or 2).
PPSD	Phase PSD.
$S_\varphi(f_m)$	PPSD at modulation frequency f_m, one-sided.
$S_{2,\varphi}(f_m)$	Two-sided PPSD at modulation frequency f_m.
SA	Spectrum analyzer.
seed, seed	An initial value placed in a register of an accumulator in a $\Sigma\Delta$ modulator.
sen	K_v in hertz.
$\text{sinc}(x)$	$=\sin(\pi x)/(\pi x)$.
STF	Signal transfer function, the ratio of the average output to the input in a $\Sigma\Delta$ modulator.
SQDSM	Single-quantizer delta–sigma modulator.
S&H	Sample and hold.
T_{segm}	Duration of a waveform segment.
T_{sequ}	Duration of a repeated sequence.
TDC	Time-to-digital converter.
VCO	Voltage-controlled oscillator.
δ	A small change.
$\delta_{a,b}$	Kronecker delta: $\delta_{a,b=a} = 1$, $\delta_{a,b\neq a} = 0$.
ω_x	Radian frequency; see f_x for subscripts x.
$*$	Convolution.
$\equiv$	Is identically equal to.
$\triangleq$	Is defined as.

SUBSCRIPTS

b   Binary: 10_b is two in binary.

d   Decimal: 10_d is 10 in decimal.

segm   Segment, referring to the part of a waveform analyzed.

sequ   Sequence, referring to the repeated digital sequence from the $\Sigma\Delta$ modulator.

CHAPTER 1

INTRODUCTION

This book is a continuation of *Frequency Synthesis by Phase Lock*, 2nd edition (*FS2*) [Egan, 2000] with significant emphasis on the study of sigma–delta ($\Sigma\Delta$) frequency synthesis. Although $\Sigma\Delta$ synthesis has already been introduced in *FS2* (Section 8.3), there is much more to learn, so much that there is a danger of the reader being overwhelmed. For this reason, we will proceed with experimental observation, one experiment at a time, picking up information as we go. We will depend on *FS2* to define a level of knowledge about frequency synthesis that precedes this new study. It will provide a ready reference for background, but we will provide enough information to make referral unnecessary, albeit helpful, for the reader who has acquired basic knowledge about frequency synthesis elsewhere.

The experimental observations will be based on Simulink® simulations. While it is not necessary to use the Simulink program to follow the results that will be discussed, models and discussions will be subsequently provided to enable the reader to perform the Simulink simulations and to progress from there to new simulations to answer new questions. Some books and papers are largely based on development of one or more ICs or delve into monolithic circuit realizations. These have advantages, but the use of simulations permits easier demonstration of a large variety of different configurations and effects.

Following our discussion of $\Sigma\Delta$ synthesis, we will consider two topics that are important when simultaneous wide bandwidth and small frequency steps are to be

Advanced Frequency Synthesis by Phase Lock, First Edition. William F. Egan.
© 2011 John Wiley & Sons, Inc. Published 2011 by John Wiley & Sons, Inc.

achieved without the use of $\Sigma\Delta$ synthesis. Finally, we will introduce an area of current intense interest, all-digital frequency synthesis. This has become important because the deep submicron CMOS technology that is being used to achieve advances in digital circuits is increasingly incompatible with analog circuits.

In order not to interrupt the main flow, many developments are included in Appendices. This also allows the information to be easily accessed from multiple places in the main text. Each appendix is designated by a letter, and an attempt has been made to choose a letter that could be associated with the content (e.g., L for loop, N for noise) as an aid to recalling the location of the material. Equations, figures, sections, and so on in Appendix Z are numbered Z.n, where Z is the letter designation for that appendix [e.g., Eq. (C.3.2) for an equation in Appendix C]. Such items in *FS2* are referred to by a designator F.n, as if *FS2* were an appendix, but the corresponding designator in *FS2* is just n [e.g., Eq. (F.4.5) refers to Eq. (4.5) in *FS2*]. A more detailed explanation is given in Appendix F.

1.1 PHASE-LOCKED SYNTHESIZER

The basic phase-locked frequency synthesizer is illustrated in Fig. 1.1a. The reference for the loop is a source at a fixed frequency f_{ref}. It is commonly derived by frequency division from a fundamental reference at frequency f'_{ref}. The phase of the fixed signal at f_{ref} is compared with the phase of the signal at frequency f_{out}/N from the frequency

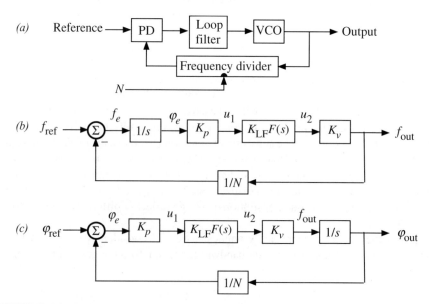

FIGURE 1.1 Basic phase-locked frequency synthesizer: (*a*) function block diagram and (*b* and *c*) mathematical block diagrams.

divider, where f_{out} is the frequency of the VCO and the loop output and N is the divider ratio. The phase comparison occurs in the phase detector (PD). The output of the PD is passed through the loop filter to drive the VCO, completing the loop.

Figure 1.1b is the mathematical block diagram representing this process. An integration is required to convert the frequency difference between the reference and divider output to a phase difference, so the Laplace representation for integration, $1/s$, must be placed in the loop. It can be placed after the frequency difference, as in Fig. 1.1b, or after the VCO, as in Fig. 1.1c. The representation chosen depends on whether we wish to consider frequency or phase at the terminals of the synthesizer. We will see that the representation of the all-digital phase-locked synthesizer differs in one significant way from Fig. 1.1.

1.2 FRACTIONAL-N FREQUENCY SYNTHESIS

The steady-state value of the frequency from a locked loop is

$$F_{out} = F_{ref}N. \tag{1.1}$$

The ratio N is inherently a whole number. Sometimes it is advantageous to employ a value of N that contains a fraction,

$$\overline{N} = N_{int} + n_{fract}. \tag{1.2}$$

To do this, we can let $N = N_{int}$ sometimes and $N = N_{int} + 1$ other times, so the average value $\overline{N}$ is given by Eq. (1.2). This is called fractional-N synthesis. We can also use $\Sigma\Delta$ synthesis to obtain improved noise characteristics through the use of other particular sequences of N-values that average to $\overline{N}$.

1.3 REPRESENTING A CHANGE IN DIVIDE NUMBER

The most common transient expected in frequency synthesis results from a change in divider ratio N. We need a way to represent this transient, which results from a change *to a loop parameter*, in a manner that permits us to obtain the loop response by analysis of a linear time-invariant circuit.

Figure 1.2a illustrates a change in N, which results in a transient at the output of the summer (e). The frequency there changes from $f_{out}(0)/N_1$ before switching to $f_{out}(0)/N_2$ after switching, where $f_{out}(0)$ is the output frequency at the moment of switching. The same transient could be caused by a frequency change, injected into the time-invariant loop shown in Fig. 1.2b, of

$$\Delta f_d(0) = f_{out}(0)\left(\frac{N_2}{N_1} - 1\right). \tag{1.3}$$

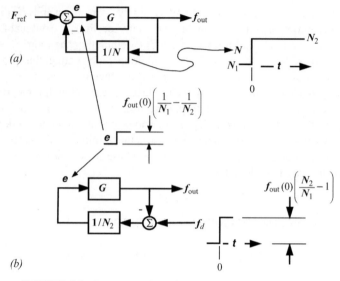

FIGURE 1.2 Representation of a change in divide number N.

If the loop is initially at steady state,

$$f_{\text{out}}(0) = N_1 F_{\text{ref}} \tag{1.4}$$

and Eq. (1.3) becomes

$$\Delta f_d(0) = F_{\text{ref}}(N_2 - N_1) = F_{\text{ref}}\Delta N, \tag{1.5}$$

which is also the eventual size of the output frequency change. Since $N = N_2$ after switching, the loop response is analyzed using that value of N. This is further illustrated in Fig. F.2.19. Thus, we can analyze the response to a change in the parameter N as if it were the response of a time-invariant circuit to a step signal at f_d, so long as the loop is initially at steady state.

During $\Sigma\Delta$ modulation, the parameter N undergoes continual changes but the output is relatively steady at the synthesized frequency,

$$f_{\text{out}}(0) = \overline{N}F_{\text{ref}}, \tag{1.6}$$

so Eq. (1.3) then becomes

$$\Delta f_d(0) = F_{\text{ref}}\overline{N}\left(\frac{N_2}{N_1} - 1\right) = F_{\text{ref}}\frac{\overline{N}}{N_1}\Delta N \tag{1.7}$$

[compare with Eq. (1.5)]. Since this now applies to the individual steps at each reference period, we can also write

$$f_d(nT_{\text{ref}}) = F_{\text{ref}}\frac{\overline{N}}{N_1}N_{\text{mod}}(nT_{\text{ref}}), \tag{1.8}$$

where N_{mod} is the modulated divider ratio.

Looking at the various frequency components in the sequence of N_{mod} values, the low-frequency components will be affected by the average value of N_1 (i.e., $\overline{N}$); so, for $f_m \ll f_{\text{ref}}$, Eq. (1.8) suggests

$$f_d(f_m \ll f_{\text{ref}}) \approx F_{\text{ref}}N_{\text{mod}}(f_m \ll f_{\text{ref}}), \tag{1.9}$$

implying that N_{mod} can be represented as an equivalent frequency deviation injected at f_d in Fig. 1.2b. However, such averaging cannot be applied to the higher frequency components. Thus, we can expect the response for $f_m \ll f_{\text{ref}}$ to be as if $f = f_{\text{ref}}N_{\text{mod}}$ were injected at the divider input [Eq. (F.8.75)], but the exact response, including the higher modulation frequencies, usually those beyond the loop bandwidth, will differ. That response is developed in Appendix Q (and its form supports the treatment of the low-frequency components that we have just described).

1.4 UNITS

We will try to use units everywhere, which may lead to some unfamiliar looking expressions. For example, rather than writing $f = 1/T$, we relate frequency f to period T by

$$f = \text{cycle}/T. \tag{1.10}$$

However, we drop or add radian units, and only radian units, at will (see Section F.1.1.4 for an explanation).

1.5 REPRESENTING PHASE NOISE

Sinusoidal frequency modulation of a signal at frequency f_m with a peak deviation Δf produces a sinusoidal phase deviation at the same frequency with a peak phase deviation of m, called the modulation index,

$$m(f_m) = \Delta f(f_m)/f_m, \tag{1.11}$$

in radians (Section F.3.1). This causes sidebands on the signal, spaced at multiples of f_m from the central spectral line. When m is small relative to 1 rad, the central spectral line is reduced little from its level without modulation, the sidebands fall off rapidly, and the first sideband has an amplitude of $m/2$ relative to the central line. If there are many modulating frequencies and if they are spaced very closely, or continuous, we can treat them as a phase power spectral density (PPSD) S_φ with units of rad^2/Hz

(Section F.3.7). The units of

$$S_\varphi|_{dB} = 10\,dB\,\log_{10} S_\varphi \qquad (1.12)$$

are decibels relative to a radian squared per hertz, dBr/Hz.

This relationship between the first sidebands and the discrete phase deviation implies a similar relationship between $S_\varphi(f_m)$ and the relative single-sideband power spectral density (PSD) $L_\varphi(\Delta f)$ at an offset of $\Delta f = \pm f_m$ from spectral center (which we see when observing the signal on a spectrum analyzer), requiring still that m be small, in which case (Section F.3.7)

$$L_\varphi(\Delta f = \pm f_m) = S_\varphi(f_m)/2. \qquad (1.13)$$

If the restriction on m is not met, Eq. (1.13) *is not valid*. However, by definition,

$$\mathcal{L}_\varphi(\Delta f = f_m) \triangleq S_\varphi(f_m)/2. \qquad (F.3.41)\,(1.14)$$

The density $\mathcal{L}_\varphi$ is a popular measure of oscillator phase noise because at sufficiently large $\Delta f = f_m$, the modulation index *is* small, so $\mathcal{L}_\varphi \approx L_\varphi$ at those offsets. Thus, $\mathcal{L}_\varphi$ indicates what the power spectrum of a signal looks like, except close to the center, where S_φ (and thus $\mathcal{L}_\varphi$) continue to climb with decreasing Δf, even as L_φ approaches a peak at $\Delta f = 0$. As $f_m \Rightarrow 0$, the measurement of S_φ requires ever narrower filters and takes ever longer times, causing a practical limit on the minimum f_m at which it is measured, but no such limit exists for L_φ.

The density $L_\varphi(\Delta f = \pm f_m)$ is called single-sideband density because it is the relative density on either side of the spectral center, $L_\varphi(\Delta f = +f_m)$ or the equivalent $L_\varphi(\Delta f = -f_m)$. We will not use double-sideband density.

The alternative use of $\mathcal{L}_\varphi$ or S_φ should present no problem because of the simple relationship Eq. (1.14), which can also be expressed as

$$\mathcal{L}_\varphi|_{dBc/Hz} \equiv S_\varphi|_{dBr/Hz} - 3\,dB. \qquad (1.15)$$

In the Fourier domain, positive and negative frequencies are used, and the power densities are divided evenly between the positive and negative frequencies. We will generally use one-sided densities and will designate two-sided (Fourier) densities by using a subscript 2. Thus,

$$S_{2,\varphi}(f_m) = S_\varphi(f_m)/2 \qquad (1.16)$$

and

$$\mathcal{L}_\varphi(\Delta f = f_m) = S_{2,\varphi}(f_m). \qquad (1.17)$$

A relationship similar to Eq. (1.16) holds also for power spectral density, but $L_\varphi(\Delta f = \pm f_m)$ is the *relative* PSD, normalized to the signal power, so it is the same one-sided as two-sided, since both the signal power and PSD are changed by the same factor of 2.

1.6 PHASE NOISE AT THE SYNTHESIZER OUTPUT

The various noise sources in the synthesizer and their effects on the output spectrum are covered extensively in *FS2*, especially in Chapters 3 (density) and 5 (discrete). We will further consider some of these and some additional noise source in Chapter 4, in particular as they relate to synthesizer ICs and to $\Sigma\Delta$ synthesizers.

1.7 OBSERVING THE OUTPUT SPECTRUM

We will observe the effects of fractional-N synthesis on the synthesizer's output spectrum using a Simulink model of a type-2 PLL with a 58 kHz unity gain bandwidth f_L. See Section L.1 for details. We will initially use a sample-and-hold (S&H) phase detector (Appendix P) to make the phase values easier to observe. Eventually, we will change to the more common phase frequency detector (PFD).

CHAPTER 2

FRACTIONAL-N AND BASIC $\Sigma\Delta$ SYNTHESIZERS

Following are some of the reasons for using a fractional-N synthesizer (Section F.8.3):

- to obtain fine resolution (step size) while maintaining a wide bandwidth for switching speed or suppression of VCO phase noise;
- to obtain fine resolution without the analog circuitry required for multiple loops;
- to minimize the divide ratio N in order to not multiply the reference phase noise by a larger number; and
- to obtain rapid phase-continuous switching for frequency sweeping.

In the fractional-N synthesizer, the divide number is changed periodically; so the average synthesized frequency is between values that could be obtained with integer N. The sequence of divide numbers is typically controlled by an accumulator that periodically produces an output to alter the value of N. Here is a simple example.

2.1 FIRST-ORDER FRACTIONAL-N

Suppose that the effective value of N is to be 10.125, that is, 10 and 1/8. Then N will be 11 for 1 out of 8 reference periods and 10 for the other 7. This can be done by accumulating the number 1/8—adding it to an existing value each reference period—and using the carry (overflow) output to increase N from 10 to 11. Table 2.1

Advanced Frequency Synthesis by Phase Lock, First Edition. William F. Egan.
© 2011 John Wiley & Sons, Inc. Published 2011 by John Wiley & Sons, Inc.

TABLE 2.1 Accumulator Contents with 1/8 Input

Initial	Input	Carry.Final
.001	.001	0.010
.010	.001	0.011
.011	.001	0.100
.100	.001	0.101
.101	.001	0.010
.110	.001	0.111
.111	.001	1.000
.000	.001	0.001

shows the repeated sequence (the first row occurs again after the last row) of binary fractions that results from accumulating the fraction $1/8 = 0.125_d = 0.001_b$.

The accumulator (Fig. 2.1 and Appendix C) might consist of a 3-bit register and an adder. One might feel more comfortable considering the least significant bit of the register to be 1, but considering it to be $2^{-3} = 0.001_2$ is consistent with the divide number N. The carry is then 1 and the accumulator input is a binary fraction. Then we need not be concerned with the number of bits in the register, as long as there are enough to hold the bits in the input word. Three bits are adequate for a 0.001_2 input but 20 bits work as well. In either case, carry (overflow) occurs at an accumulated value of 1.

In Table 2.1, we see a carry output for one of the eight periods. Assuming that the synthesizer's output frequency is maintained constant (e.g., due to a narrow loop) during this sequence, it will have the value $10.125f_{\text{ref}}$, corresponding to the average value $\overline{N}$. The frequency at the output of the divider will be $10.125f_{\text{ref}}/10$, exceeding f_{ref}, during most cycles, but it will be lower than f_{ref}, at $10.125f_{\text{ref}}/11$, every eighth cycle. This will produce a sawtooth-like output from the PD in which the phase decreases for seven cycles and increases at the eighth. Note how the number in the accumulator in Table 2.1 ("Initial" or "Final") does just the opposite, increasing continually for seven periods and decreasing for the eighth. This value can be used to cancel the PD waveform that results from the fractional division. Otherwise, the loop would have to be made narrow enough to attenuate the resulting phase modulation sidebands on the output spectrum.

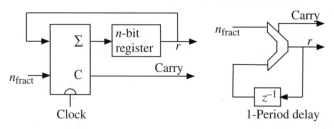

FIGURE 2.1   Two representations of the accumulator.

TABLE 2.2 **Accumulator Contents with 3/8 Input**

Initial	Input	Carry.Final
.000	.011	0.011
.011	.011	0.110
.110	.011	1.001
.001	.011	0.100
.100	.011	0.111
.111	.011	1.010
.010	.011	0.101
.101	.011	1.000

If the average value $\overline{N}$ were to be 100.375, the repeated sequence would be modified as shown in Table 2.2.

We can see a carry output for three of the eight periods. As before, the length of the repeated sequence is 2^n, where n is the number of stages required in the register ($2^3 = 8$ here). If we change the fraction to $n_{fract} = N_{fract}/2^3$, where N_{fract} is some even number (N_{fract} is 1 and 3 in Tables 2.1 and 2.2, respectively), the sequence will be shorter. With $N_{fract} = 2$, for example, it could be 0.00, 0.01, 0.10, 0.11. Even if we start with 0.001, the sequence will still be only four periods long (i.e., 0.001, 0.011, 0.101, 0.111).

We can also add any number of stages beyond the number required for the input (unless we preset a smaller value in the register) without changing the length of the sequence, which depends on the least significant digit in the input, not counting any unnecessary trailing zeros.

The block diagram in Fig. 2.2 illustrates the concept of a $\Sigma\Delta$ synthesizer. For our first-order fractional-N synthesizer, the $\Sigma\Delta$ modulator is an accumulator, which is a first-order $\Sigma\Delta$ modulator. The summation shown here is implemented by a pulse swallower (Section F.4.4) in the frequency divider.

2.1.1 Canceling Quantization Noise

Figure 2.3 shows waveforms when the fractional value is $1/16 = 0.0625$. The carry output is at the bottom. It is labeled "swallow" because it would be the swallow command to a pulse swallower that would increase the value of N by 1. The PD output (equal to phase in cycles, because of the value chosen for K_p) is at the top. The phase

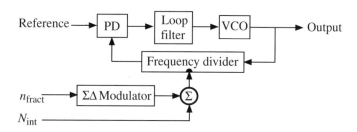

FIGURE 2.2 Conceptual block diagram of a $\Sigma\Delta$ synthesizer.

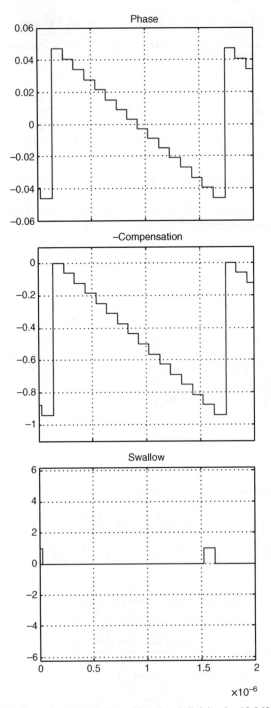

FIGURE 2.3 Waveforms from simulation of fractional division by 10.0625. *Top*: PD output (phase in cycles). *Middle*: Negative of compensating waveform before gain (accumulator contents). *Bottom*: Swallow command to divider. The reference frequency is 10 MHz.

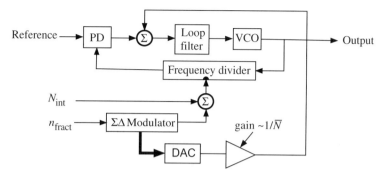

FIGURE 2.4 Conceptual block diagram of a $\Sigma\Delta$ synthesizer with analog cancellation of the PD output waveform.

steps down in 15 equal increments and then steps up to the initial value. The last step equals 1 down increment plus 16 up increments. The 16-increment phase increase is produced by one additional (swallowed) input cycle to the frequency divider and corresponds to a phase change, at the PD, of one cycle divided by $\overline{N} = 10.0625$.

In the middle is the accumulator's register content, inverted to make it easier to compare with the PD output. Note how the noninverted register value (after conversion to a voltage or current) could be added to the PD output to cancel the PD output waveform, the quantization noise. The concept is illustrated in Fig. 2.4 (see also Fig. F.8.17). The fraction in the accumulator would be multiplied by $K_p/\overline{N}$ in a DA converter because the range of the cancellation (1) represents a one-cycle change at the divider input and that is equivalent to $K_p\,\text{cycle}/\overline{N}$ at the PD output (see Section F.6.A). Appendix C gives a detailed mathematical description of the relationship between accumulator contents and the PD output waveform.

It is also possible to drive a phase or delay modulator, at the frequency divider output, with the same compensation signal, thus canceling the phase modulation rather than the resulting voltage [Banerjee, 2008]. If the modulator is placed at the frequency divider input, the required phase modulation does not depend on $\overline{N}$, but time delay modulation would. We will not consider cancellation by phase modulation further.

2.1.2 Cancellation with a PFD

Figure 2.3 is taken from a simulation of a synthesizer loop using a sample-and-hold (S&H) PD (Appendix P and Section F.5.6) in order to show the equivalence between the phase variation and the number in the accumulator. Very accurate cancellation is more difficult to obtain with the more common charge pump PD because of its pulse output waveform. Nevertheless, one can add its PD pulse to a pulse whose area is proportional to the register contents and sample and hold (resample) the sum after they have canceled. This technique is illustrated in Figs. 2.5 and 2.6 [Liu and Li, 2005]. The sampling has to occur after the integrator capacitor. Otherwise, the pulses would be gone and there would be nothing to sample (see Section H.2). We have added an inhibit input that can be used to keep the switch closed in order to prevent a possible problem during frequency switching, which is discussed in Appendix H.

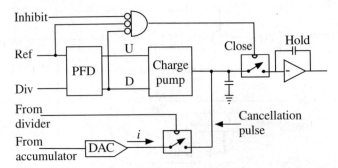

FIGURE 2.5 Sampled cancellation circuit (with added inhibit to allow the switch to be kept closed).

This also provides the opportunity to make the cancellation proportional to $1/N$ by generating the cancellation pulse as a fixed number of cycles of the input frequency. If some fixed number N_x of cycles, instead of all N, are used to generate the cancellation pulse, the number by which the accumulator fraction must be multiplied

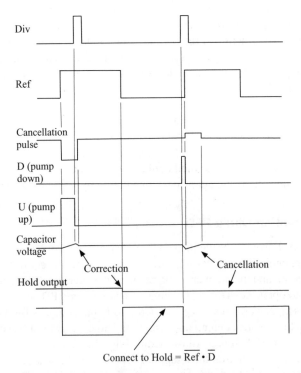

$$\text{Connect to Hold} = \overline{\text{Ref} \cdot \overline{D}}$$

FIGURE 2.6 Waveforms for sampled cancellation. The Hold output (which is inverted) responds to a required PD correction, but not to the transient caused by the canceling waveforms.

must be increased by $\overline{N}/N_x$, giving a required multiplication of

$$\frac{K_p \text{ cycle } \overline{N}}{\overline{N}} \frac{\overline{N}}{N_x} = \frac{K_p \text{ cycle}}{N_x}, \tag{2.1}$$

but this is a constant, something that does not change with the frequency being synthesized.

The accumulator fraction (or the current it produces) might be shifted by a constant (or AC coupled) to give it zero average value. Otherwise, the loop will generate a fixed phase error to cancel its average value. This is of little consequence with a S&H PD, but we usually like to operate a PFD near zero phase to reduce the static pulse width.

Although the circuit of Fig. 2.5 improves the cancellation, significant cancellation can be obtained even without it (i.e., outputting just the unsampled capacitor voltage). The first synthesizer of this type (Section F.8.3.1) used a cancellation pulse having a width proportional to the output period in canceling a phase detector output that was approximately a square wave.

However, in the absence of resampling, attempts to precisely cancel the PFD pulse are hampered by the variation in the time of occurrence of that pulse relative to the reference. Meninger and Perrott [2003] used a lookup table, addressed by the accumulator contents, to modify the input to the DAC that provides the correction signal in order to compensate, approximately, for the varying sample time at the PFD. This reduced spurs at an offset of f_{fract} by up to 22 dB in simulations of a first-order fractional-N synthesizer.

2.1.3 Cancellation Techniques

A number of techniques have been used to improve cancellation. These include

- use of $\Sigma\Delta$ modulators to reduce the number of bits driving the DAC,[1,2]
- use of mismatch error cancellation techniques [Schreier and Temes, 2005] to reduce the detrimental effects of nonlinearity in DACs,[1,2]
- use of the same currents for cancellation that are used for the PD pulses,[1]
- random interchanging of current sources,[1] and
- careful design to minimize timing errors.[1]

Here is a cursory summary of some results from the use of these techniques:

- First-order fractional-N, simulation[1]:
  - broadband quantization noise: 36 dB reduction
  - fractional spurs: < -90 dBc
- Second-order $\Sigma\Delta$ synthesizer with LSB dither, experiment[2]:
  - broadband quantization noise: 20 dB reduction
  - fractional spurs: ≥ 8 dB reduction

2.1.4 Spectrum Without Cancellation

The FM sidebands caused by the quantization noise can be seen in Fig. 2.7. The central spectral line at 100.625 MHz has been converted to 4 MHz for observation. The amplitude of this discrete line is obtained by multiplying the indicated density by the noise bandwidth. This gives 0 dB. The same level is obtained with no modulation, consistent with small modulation index. See Appendix S for a discussion of spectrum analysis.

The first sidebands are at a relative level close to the attenuation of the loop [obtain from Fig. L.4*b* or execute MATLAB® script `ct58at.m`, type `ct58at(625e3)`], −34.8 dB at 625 kHz, as predicted in Section F.8.3.3 for an approximate sawtooth phase error signal. The sideband that occurs when n_{fract} is $1/L_a$ or $(L_a - 1)/L_a$, where L_a is the length of (number of states in) the accumulator, is sometimes called an integer boundary spur. For $n_{\text{fract}} = 0.0625$, L_a is effectively 16, even if there are unused LSBs. If we change the fraction to 3/16, we get 3 stepped triangles over 16 periods, as occurred in Table 2.2, and the first sidebands are reduced by about 13 dB.[3] The time for the pattern to repeat is 16 reference periods, so spurious products are generated at multiples of 1/16th of the reference frequency. One-sixteenth of the reference frequency also equals the frequency increments that can be produced by this process, so it is called the channel spacing.

If we start with some initial value in the accumulator, perhaps a very small one, an additional offset will be added to the register contents relative to the phase detector output, but the shape of the cancellation will not be affected nor will the repetition interval. While it has been argued (Section F.8.3.4.3) that an initial setting can cause the sequence to go through a different set of values, one of which may produce a better

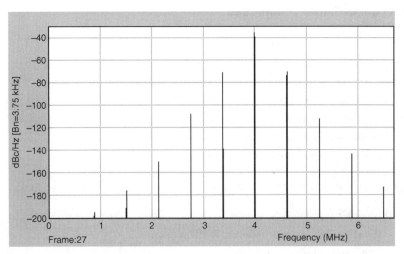

FIGURE 2.7 Output spectrum due to quantization noise, 100.625 MHz converted to 4 MHz. The *y*-axis is calibrated for noise density, but the scale is still useful for obtaining relative sideband levels of discrete spurs. S&H PD. Sample rate varies (the significance of this will be discussed).

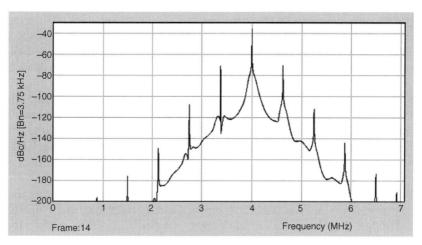

FIGURE 2.8   Spectrum of output at 100.62515... MHz (after conversion to 4 MHz). Lesser spectral lines are spaced by about 152 Hz, causing them to appear as noise in the 3750 Hz resolution bandwidth. S&H PD. Sample rate varies.

set of spurs, we have not observed any change in the spectrum *in the absence of cancellation* (with the single accumulator).

If we add a small fraction, 2^{-16}, to n_{fract}, making it 0.06251525..., the sequence requires $2^{16} = 65{,}536$ periods of f_{ref} to repeat [Hosseini and Kennedy, 2006], an increase of $2^{12} = 4096$. Therefore, there will be 4095 spectral lines between each pair in Fig. 2.7. With the resolution of Fig. 2.8, this appears as noise, but the dominant spurs still appear at about the same level as previously.

2.1.5 Influence of $\overline{N}$

What we have discussed appears to depend only on the fractional value n_{fract} and not on the integer part N_{int} of $\overline{N}$. However, the magnitude of the phase variations is greatly reduced at larger $\overline{N}$, and that reduces some nonlinear effects. For example, the center of the PD pulse varies more with a smaller value of $\overline{N}$ and this might be more detrimental to cancellation if the sampler of Fig. 2.5 is not employed. However, if we use a larger value of $\overline{N}$ for a given output frequency, f_{ref} will be proportionally lower and the spurs, which are spread over f_{ref}, will be correspondingly closer together, leading to a higher power spectral density. We will use $N_{\text{int}} = 10$, a relatively small number, to emphasize some of the effects we wish to observe.

2.2 SECOND-ORDER FRACTIONAL-N

2.2.1 Purpose

Even with the circuit of Fig. 2.5, suppression of spurs is limited by the accuracy of the DAC. King [1980] found that the PD output waveform can be manipulated to reduce its contribution of large spurs, thus making the cancellation process more effective (Section F.8.3.4.1). To this end, the contents of the accumulator are accumulated in a second accumulator. When the second accumulator reaches its capacity (i.e., 1), its carry is used to increase N by 1. This is the same thing that occurs when the first accumulator generates a carry. However, during the following period of F_{ref}, N is reduced by 1. Because of this countering action, there is no effect on the average value of N. However, a phase pulse of one input cycle is generated, producing a net integrated phase of zero from the sum of that pulse and the phase corresponding to the contents of the first accumulator that instigated it (Fig. F.8.22). In other words, whenever the value at the PD output reaches the equivalent of one input cycle, a countering phase pulse reduces it to zero.

When performing a Fourier series analysis to determine the magnitude of the spurs, one would multiply the phase error by a sinusoid with a period equal to the repetition period of the waveform, or a harmonic thereof, and then integrate the product. If the integral of the phase error is kept to a low value, the integral after multiplication by a sinusoid at the fundamental frequency, or a small multiple thereof, will also tend to be small, corresponding to smaller spurs.

2.2.2 Form

Figure 2.9 illustrates the simple fractional-N configuration that we have just studied, while Fig. 2.10 shows the second-order fractional-N configuration. Note that the value of N in Fig. 2.10 can increase by as much as 2 or can decrease by 1 relative to N_{int}. Note also that the cancellation waveform comes only from the second accumulator and is processed just as is the carry from that accumulator (analysis in Appendix C).[4] We have created a sequence of N values that will produce generally weaker spurs and the sequence to cancel those weaker spurs. Hopefully, even with inaccuracies in the cancellation, we will have a cleaner spectrum than if the spurs had not been initially reduced.

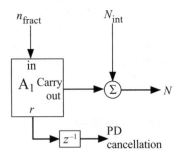

FIGURE 2.9 Generation of N for fractional-N synthesizer. Block A represents an accumulator (Fig. 2.1) and r is its register contents. The delay (z^{-1}) is for proper timing.

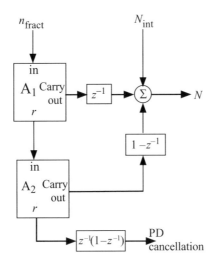

FIGURE 2.10   Generation of N for second-order fractional-N. The delayed negative output is indicated by $-z^{-1}$.

2.2.3 Performance

The PD output, the inverse of the PD cancellation from Fig. 2.10, and the combined carry output from Fig. 2.10 are shown in Fig. 2.11, while Fig. 2.12 shows the spectrum (without cancellation). The spectral lines at ±625 kHz have been reduced by about 18 dB, but there are lines at 312.5 kHz intervals. The lines at ±312.5 kHz are 44 dB below the central line; 22 dB of that is due to the loop response. This is 22 dB lower, excepting the loop attenuation, than were the first sidebands with only one accumulator. The 312.5 kHz spacing suggests that 1 cycle/312.5 kHz = 3.2 μs is the repetition period of the PD waveform in Fig. 2.11. Because these spurs are spaced at less than the channel spacing, the spacing that occurs with a first-order fractional-N circuit, they are sometimes called subfractional spurs.

Figure 2.13 shows the effect of the same small frequency offset used in Fig. 2.8 on the spectrum. This time the spectrum is more noise-like, the spurs spaced at 312.5 kHz increments have disappeared below the noise, and the ±625 kHz sidebands are lower by a few decibels than they were without the frequency offset.

For the first-order fractional-N, an initial value in the accumulator made no difference in repetition interval or the observed spectrum; but for this second-order fractional-N, an initial value can increase the sequence length. With an initial value 2^{-n} in the first accumulator (making it effectively an n-bit accumulator), the sequence length for this second-order modulator is between 2^{n-1} and 2^{n+1}, between 32,768 and 131,072, when $n = 16$ (see Section X.2). Thus, the number of spectral lines in 10 MHz is at least 32,768, implying a spacing of 305.1... Hz, much smaller than the resolution bandwidth in Fig. 2.14; so the lines appear as a continuous spectrum between the major spurs. Whether the lines appear individually or as

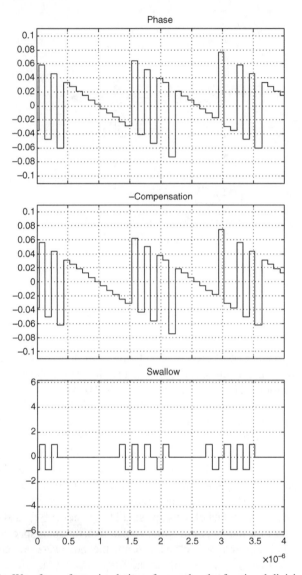

FIGURE 2.11 Waveforms from simulation of second-order fractional division by 10.0625. *Top*: PD output (phase in cycles). *Middle*: Negative of compensating waveform (accumulator contents after gain). *Bottom*: Swallow command to divider. The reference frequency is 10 MHz.

noise depends on the resolution bandwidth of the spectrum analysis. Whether they act upon a system as individual spurs or as noise depends on the system (see Section S.4).

Note that there is not a fixed offset in the cancellation this time, since the accumulator value is being processed by a difference operation $(1 - z^{-1})$. The initial value (seed) changes the output now because it affects the integration by the second accumulator.

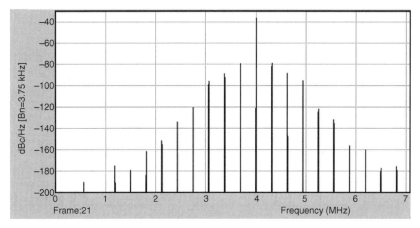

FIGURE 2.12 Output spectrum at 100.625 MHz with second-order fractional-N. S&H PD. Sample rate varies.

2.2.4 Interpreting the Spectrum

Refer also to Appendix S.

We note multiple overlapping spectrum shapes in Fig. 2.14. The buffer size for this spectrum analysis is 8192 and the simulated sample rate is 20.48 MHz (5.12 times the center frequency), so the simulated duration of the analyzed segment T_{segm} is 400 μs.[5] The length of the $\Sigma\Delta$ sequence is between 2^{15} and 2^{17} and it occurs at a simulated rate of $f_{ref} = 10$ MHz, requiring a period T_{sequ} between 3.2768 and 13.1072 ms. Thus, each FFT covers only about 3–12% of the sequence. Once the synthesizer is settled, variations in the output phase will have the same period as the sequence, so subsequent FFTs are transforms of different portions of the waveform.

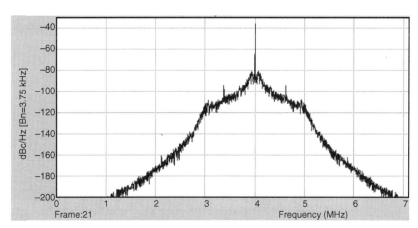

FIGURE 2.13 Output spectrum at 100.62515... with second-order fractional-N. The roll-off at ±1 MHz is due to a sharp filter within the loop. Its main purposes are to prevent fold-over of the observed spectrum due to mixing or aliasing. S&H PD. Sample rate varies.

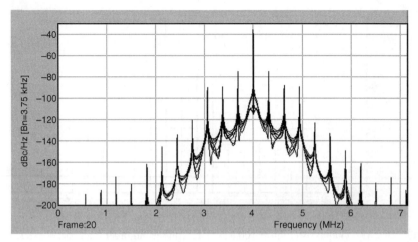

FIGURE 2.14   Output spectrum at 100.625 MHz with second-order fractional-N and 2^{-16} initial value in the first accumulator. Ten frames superimposed. FFT: 2^{13} (8192) buffer size, 20.48 MHz sampling. S&H PD. Sample rate varies.

Variations at a basic frequency of 312.5 kHz will appear in each T_{segm}, so all will have a PSD with sidebands offset by multiples of 312.5 kHz, but variations with periods longer than T_{segm} will cause differences in the details of each frame.

Figure 2.15 shows the PSD using a buffer that is 32 times larger and a sample rate of 20 MHz (which causes a slight shift in center frequency[6]), producing a transform that represents a waveform that repeats with a period equal to the maximum sequence length, just as the output signal will. However, each frame[7] required about 26 min of

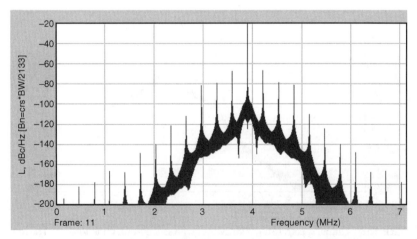

FIGURE 2.15   Output spectrum under the same conditions as in Fig. 2.14 except FFT has 2^{18} buffer size, 20 MHz sampling. Noise bandwidth is 114 Hz. The central line is at −20.6 dB.

simulation (real) time for Fig. 2.15, whereas a frame took less than 45 s for Fig. 2.14. While an analysis of duration equal to the repetition interval is required for an accurate spectral analysis (Fourier series), the simulation time required can lead us to use shorter buffers. We begin to treat the spectrums as though they were noise spectrums with random variations from frame to frame rather than what they are, Fourier transforms of long deterministic sequences with deterministic variations between frames.

It appears that the sequence length here is the maximum, $2^{n+1} = 2^{17}$. This was determined by experiments with a slightly shorter register where $n = 14$ (to save time). A display like Fig. 2.15 was obtained when the FFT sample duration T_{segm} was equal to the maximum sequence duration T_{sequ}. When the length of the FFT was doubled, the space below the spectrum filled in completely; there was no white space. This suggests the presence of cells containing little power, consistent with an FFT duration equal to twice the repetition interval T_{sequ}, wherein alternate responses are ideally zero (all the spectral lines are at harmonics of cycle/T_{sequ}).

The noise bandwidth B_n depends on the buffer size.[8] Even if it were narrower than the spacing of the minor spurs (that look like noise) in Fig. 2.15, resolution would depend on the minimum linewidth of the display; each megahertz in Fig. 2.15 contains more than 13,000 B_n. B_n also affects the apparent amplitude of discrete spurs (the central line has amplitude -10 dB $\log_{10} B_n$ as a consequence of the scaling in terms of dBc/Hz), but not their values relative to the central line. (See Appendix S for further discussion of spectral displays.)

Note also that the carry now varies between -1 and $+2$, as can be seen for $n_{\text{fract}} = 0.5625$ in Fig. 2.16. This reflects the possible carry combinations in Fig. 2.10.

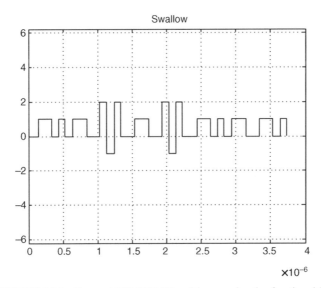

FIGURE 2.16   Carry at 105.625 MHz with second-order fractional-N.

2.3 HIGHER ORDER FRACTIONAL-N

Figure 2.17 shows a fourth-order fractional-N circuit with cancellation and is indicative of the circuit for higher and lower orders (see Appendix C). This multistage configuration is called MASH-1111,[9] indicating that it is a ΣΔ modulator consisting of four first-order error feedback modulators (EFMs) or single-quantizer ΔΣ modulators (SQDSMs), each producing a binary (two-level) output, such as we have been studying.[10] Higher order EFMs will be considered later.

Figure 2.18 shows the output spectrum of a third-order fractional-N synthesizer (MASH-111) with an initial value (seed) of 2^{-16} in the accumulator. We expect this combination to produce a noise-like spectrum free of discrete spurs [Kozak and Kale, 2004]. However, Fig. 2.18 shows discrete spurs at ± 312.5 and ± 937.5 kHz. Moreover, the phase power spectral density (one sided) for a pth-order MASH is expected to be (Appendix M)

$$S_{\varphi,\text{out}} = S_\varphi |H(f_m)|^2, \tag{2.2}$$

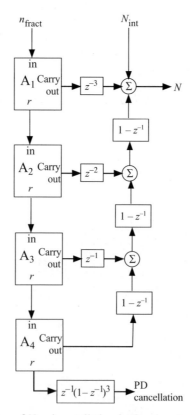

FIGURE 2.17 Generation of N and cancellation for fourth-order fractional-N. This shows a MASH-1111 architecture and indicates the pattern for multiple accumulator stages.

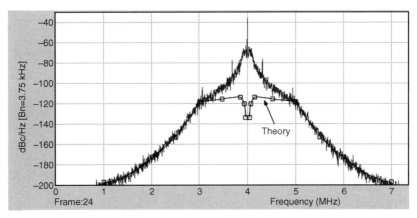

FIGURE 2.18 Output spectrum at 100.625 MHz with third-order fractional-N and 2^{-16} initial value in the first accumulator, S&H PD. Theoretical curve is also shown. Sample rate varies.

where

$$S_\varphi = \frac{(2\pi\ \text{rad})^2}{6 f_{\text{ref}}} \left[2 \sin\left(\pi \frac{f_m}{f_{\text{ref}}} \right) \right]^{2(p-1)}, \qquad (2.3)$$

and $p = 3$ for this modulator, but the corresponding sideband density (Section L.1.4), as shown in Fig. 2.18, is far different from the observed density. We might suspect that the phase variance is too large for the small-modulation-index relationship between S_φ and L_φ to pertain (Section F.3.7), but this is not the case. (Note that the large variance problem is common, close to the carrier due to oscillator noise, but the only noise that we have included in our analysis is the quantization noise.) The PD waveforms in Fig. 2.19 provide a clue. Equation (2.3) was developed from z-transforms, implying a constant sampling frequency, but Fig. 2.19 clearly shows that the sampling frequency is not constant; the first positive pulse (upper trace) is 80 ns wide and is followed by a 119 ns negative pulse.

Note, however, that if the accumulator is synchronized with the divider output, it will possess the same timing variations as the sampled phase, so a signal derived from the accumulator can still be used to cancel the output waveform from a S&H PD.

2.3.1 Constant Sampling Rate

Let us make the sampling rate constant by switching the inputs to the PD,[11] and reverse the sign of K_p by giving the ramp a negative slope. The first action changes the sign of the phase error and the second changes it back (Fig. 2.20). The loop mathematical diagram will remain the same, but the PD output will update in synchronism with the constant F_{ref}. There will be an additional (varying) delay from the divider output transition to the sample time, about half a cycle at F_{ref} (about $1°$ at

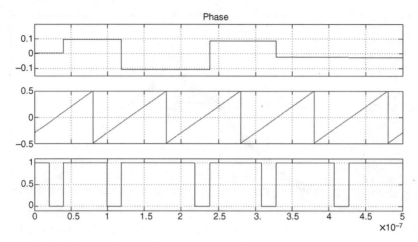

FIGURE 2.19   PD waveforms. The ramp in the middle is synchronized with the reference input. It is sampled when the divider output at the bottom goes up. The sampled value, indicating phase, is at the top.

the frequency of unity loop gain for this loop). The PD waveforms are shown in Fig. 2.21. Note that the ramp no longer has a constant amplitude; the effect is severe with a third-order accumulator and N only about 10. But we also see in Fig. 2.22 that the spectrum now has the theoretical shape.

Alternatively, we could sample and hold the output of the original (not reversed) S&H PD in synchronism with the reference, but half a cycle after the phase comparison (see Appendix P). When a PFD is employed, we can similarly sample and hold the voltage on the first integrator capacitor. Figure 2.23 shows a setup that is useful for experiments comparing performance with and without sampling. The sampler in Fig. 2.5, less the cancellation circuitry, has also been used [Cassia et al., 2003]. The frequency switching performance of synthesizers using these methods, as well as a modification of Fig. 2.5, is described in Appendix H. It is important that the S&H provide a constant sampling rate in steady state and not be detrimental to the response during frequency switching; there is usually no need for it to meet these requirements simultaneously.

The possibility that the sampled signal might not have settled by the time another sample occurs is of some concern when N is very small, as it is in our experiments. With $N = 10$, a change of ± 5 input cycles would cause a charge pulse to extend to a sample point half a reference cycle either side of the reference time. A MASH-1111

FIGURE 2.20   Mathematical diagram effectively unchanged.

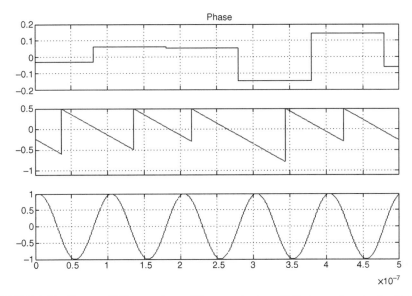

FIGURE 2.21 Reversed PD waveforms. The ramp in the middle is synchronized with the divider output. It is sampled when the reference sinusoid at the bottom goes up through zero. The sampled value, indicating phase, is at the top.

modulator produces outputs from -7 to 8. However, the phase error is determined by N after digital integration (summation) of the deviation from n_{fract}. Simulations, using the MATLAB script `mashall3.m`, indicate that the integrated value remains within the range ± 3 (see Section M.7).

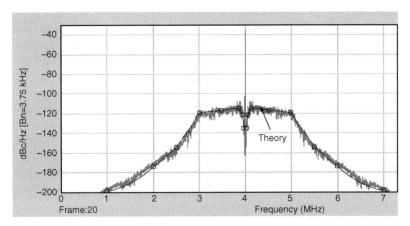

FIGURE 2.22 Output spectrum at 100.625 MHz with third-order fractional-N and 2^{-16} initial value in the first accumulator, S&H PD. Theoretical curve is also shown. Sample rate is constant.

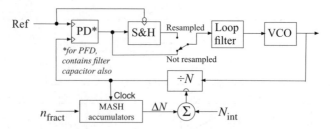

FIGURE 2.23 Experiment to compare loop with resampled PD output to one with no resampling. Sample occurs half a reference cycle after phase detection.

2.3.2 Noise Shaping Versus Cancellation

Typically, the cancellation waveform from the accumulator would not be used for higher order ($p > 2$) fractional-N circuits where we depend on ΣΔ noise shaping rather than cancellation. Besides the additional circuitry required, once noise shaping has reduced the PSD in the area of interest to a low enough value, it might be increased by inaccuracies in additional cancellation. If there would be just a gain inaccuracy, the net PD output waveform would only be reduced in amplitude and the two effects would enhance each other, but less uniform errors might destroy the noise shaping.

2.3.3 Effect of a Varying Sampling Rate

It turns out that if we do not take these measures to obtain constant sampling frequency, the effect of variations in sampling rate is less severe with the PFD than with the sample-and-hold PD.[12] Figure 2.24a shows the spectrum with a PFD without constant rate sampling, whereas Fig. 2.24b shows the spectrum with resampling. The theoretical PPSD at 0.01 MHz from spectral center is -153 dBc/Hz and the peak at about 0.15 MHz from center is -113 dBc/Hz (Appendix L). These values are close to those shown in Fig. 2.24b with constant sampling but are significantly exceeded with variable sampling in Fig. 2.24a. The conditions for Fig. 2.25b are like those of Fig. 2.24a, which is repeated as Fig. 2.25a, except that N_{int} (and thus F_{out}) is doubled and we can see that the larger value of N somewhat reduces the impact of the variable sampling rate.

Figure 2.26 shows a logarithmic plot of the spectral density from a similar loop. Data from the constant sampling frequency loop closely follow the theoretical values from Eq. (2.3) and data from the variable sampling frequency loop show additional noise that matches theory for the nonlinear effect thereby produced. The constant sampling rate data have a slope of $20(p-1)$ dB/decade, 40 dB/decade in this case where $p = 3$, and will be correspondingly steeper for higher order MASH. The other curve always has a slope of 20 dB/decade. At low frequencies, the other curve is given by (Appendix E)

$$\mathcal{L}_{out\_excess} \approx \frac{k_p}{f_{ref}\overline{N}^2}\left(\frac{f_m}{f_{ref}}\right)^2,$$

(2.4)

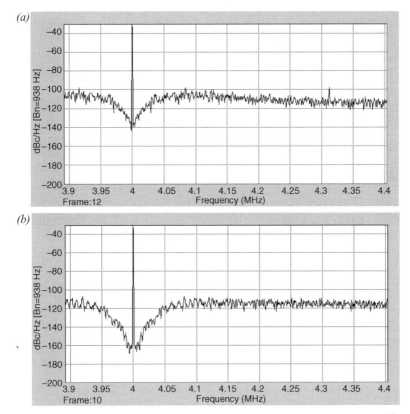

FIGURE 2.24 Output spectrum at 100.625 MHz with third-order fractional-N, 2^{-16} initial value in the first accumulator, and charge pump PD. Sample rate varies in (a), where a discrete spur is visible at 312 kHz offset. Output from integrator is resampled at a constant rate in (b). $F_{ref} = 10$ MHz.

where $k_3 = 379$, $k_4 = 5000$, and $k_5 = 69,648$. Note that while increasing the order of the MASH suppresses the close-in noise due to the higher slopes, $\mathcal{L}_{out\_excess}$ is higher for higher orders. Note that increasing N (with fixed f_{ref}) decreases the excess noise, as observed in Fig. 2.24b.

The spreadsheet `QuantN Calculator.xls` can be used to obtain $\mathcal{L}_{out\_excess}$ and $\mathcal{L}_{out\_linear}$.

Let us rewrite Eqs. (2.3) and (2.4) in a way that shows the effects of N for given values of p, f_m, and f_{out}. Equation (2.3) can be written as

$$\mathcal{L}_q|_{f_m \ll f_{ref}} \approx \overline{N}^{2p-1} \frac{(2\pi)^{2p}}{12 f_{out}} \left[\frac{f_m}{f_{out}} \right]^{2(p-1)}, \qquad (2.5)$$

where we use q because we are representing the standard quantization noise, and Eq. (2.4) can be written as

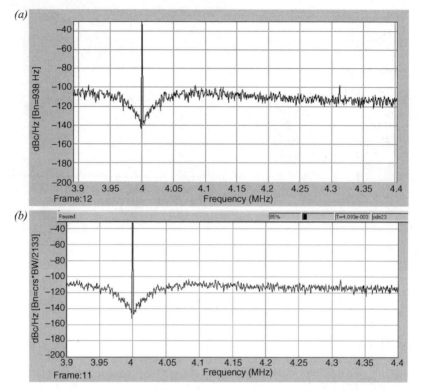

FIGURE 2.25   Spectrum in (a) is same as in Fig. 2.24a. Spectrum in (b) is for a synthesizer with the same conditions but N_{int} and F_{out} doubled.

$$\mathcal{L}_{\text{out\_excess}}\big|_{f_m \ll f_{\text{ref}}} \approx \frac{\overline{N}k_p}{f_{\text{out}}}\left(\frac{f_m}{f_{\text{out}}}\right)^2. \tag{2.6}$$

For the parameters that apply to Fig. 2.26, these are

$$\mathcal{L}_q \approx f_m^4\, 5.1 \times 10^{-32}\,\text{Hz}^{-5} \rightarrow 40\,\text{dB}\,\log_{10}\,(f_m/\text{Hz}) - 312.9\,\text{dBc/Hz} \tag{2.7}$$

and

$$\mathcal{L}_{\text{out\_excess}} \approx f_m^2\, 3.8 \times 10^{-21}\,\text{Hz}^{-3} \rightarrow 20\,\text{dB}\,\log_{10}\,(f_m/\text{Hz}) - 204.2\,\text{dBc/Hz} \tag{2.8}$$

which are plotted in Fig. 2.26. We see that $\mathcal{L}_{\text{out\_excess}}$ dominates below about 0.25 MHz. For $\mathcal{L}_{\text{out\_excess}}$ to be smaller than $\mathcal{L}_q$ at 0.01 MHz, $\overline{N}$ would have to be increased to about 52. However, increasing $\overline{N}$ would make both values higher; $\mathcal{L}_{\varphi}$ just increases faster. From this we can conclude that if the sample rate is not constant in the loop, the slope of the PSD at small frequency offsets is liable to be only 20 dB/decade when $\overline{N}$ is small (as is required for low PPSDs) rather than the theoretical 20 $(p-1)$ dB/decade.

100.15 MHz, 10 MHz Ref

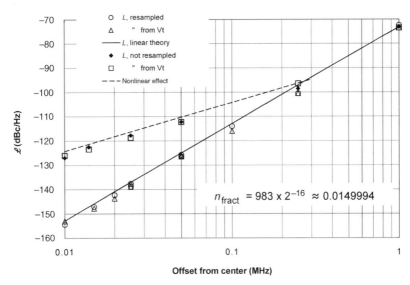

FIGURE 2.26 Single-sideband PSD for a MASH-111 loop with and without constant sampling (i.e., resampled and not resampled). Data from simulated $\mathcal{L}(f_m)$ are shown and values implied by the PSD on the tuning voltages (i.e., PPSD from V_t) plus lines showing theoretical values. Data have been divided by the theoretical loop response (unity gain is at $\sim 70\,\text{MHz}$) to show their values without the attenuation of the loop.

2.4 SPECTRUMS WITH CONSTANT SAMPLING RATE

We will now look at the spectrums from the loop that we have been using but with constant sampling rate and a charge pump phase detector.

2.4.1 100.625 MHz with Zero Initial Condition

For fractional-N loops with first- and second-order modulators, the spectrums with constant sampling rate and a charge pump PD do not differ significantly from the spectrums with varying sample rate and S&H PD (Figs. 2.7 and 2.12) under similar conditions.

For the third-order MASH-111 configuration, the spectrum is shown in Fig. 2.27. The spurs, especially those closest to the center, are smaller than they are for the second-order fractional-N but the spacing is the same. The sequence length (Section X.2) for the $0.0625 = 2^{-4}$ fraction and a third-order fractional-N is 2^5, so the transform can have spurs spaced at $2^{-5} \times f_{\text{ref}}$, that is $312.5\,\text{kHz}$. The sequence length for the second-order fractional-N is between 2^3 and 2^5, so it can have the same sequence length, depending on the fraction being synthesized, and apparently does, based on the spur spacing in Fig. 2.12.

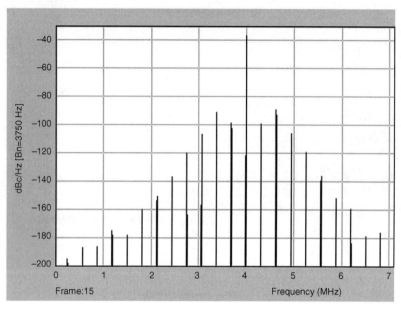

FIGURE 2.27 Output spectrum at 100.625 MHz with third-order fractional-N. Charge pump PD and integrator (CPPD&I) resampled at a constant rate. Initial zero in accumulators.

For a fourth-order fractional-N, the length can be the same as for the third-order case or it can twice as long. From Fig. 2.28, it is twice as long, giving spurs spaced at 156.25 kHz. The levels of the highest spurs are a bit lower than those for the third-order case. Note that the lines at the new frequencies are broad at the bottom. This is an artifact due to the length, 8192, of the buffer being used in generating the FFT. The central line is centered in a transform bin and the bins are spaced at 2.5 kHz.[13] Therefore, any spurs that are a multiple of 2500 Hz from the center are also centered in a bin. This is satisfied by multiples of 312.5 kHz but not for the new spurs, which occur at bin boundaries. The result is "leakage" and some small attenuation (1.4 dB) of the apparent spur amplitude (Appendix S).

So, no matter how many MASH stages we have employed, from one to four, we see discrete lines when we synthesize simple fractions, like 1/16, and start with zero initial conditions.

Here we might diverge to verify that our choice of $n_{\mathrm{fract}} = 2^{-4}$ is a representative bad case. Figure 2.29 shows the spectrum for $n_{\mathrm{fract}} = 2^{-9}$ under conditions similar to those of Fig. 2.27. Note that the noise power is spread among more spurs, appearing to be almost a continuum, and the level is lower at any offset where a spur appears in Fig. 2.27. With a first- or second-order modulator, the close-in spurs would be higher than those in Fig. 2.29. With 2^{-16} seed, the spectrum is like Fig. 2.22. We might have gone the other way and chosen a smaller fraction, 2^{-3} for example, but those higher offset spurs would be harder to see due to the attenuation of the loop (which would also hold for most synthesizer designs).

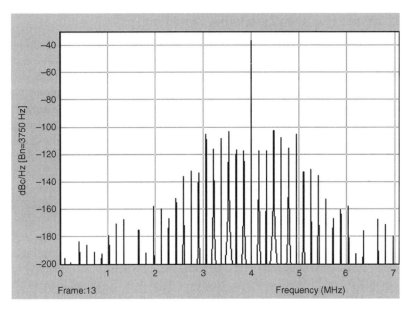

FIGURE 2.28 Output spectrum at 100.625 MHz with fourth-order fractional-N. The broadening of alternate lines is an artifact of the spectrum display and could be eliminated by doubling the buffer size. CPPD&I resampled at a constant rate, initial zero.

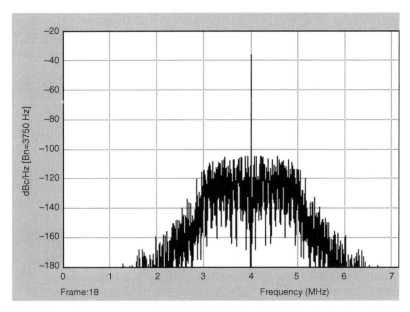

FIGURE 2.29 Output spectrum at 100.01953125 MHz with third-order fractional-N. Charge pump PD and integrator (CPPD&I) resampled at a constant rate, initial zero.

2.4.2 100.62515… with Zero Initial Condition

Miller and Conley [1991] observed (Section M.3) that a third-order fractional-N appeared to provide an output with shaped noise except for fractions that employ only the most significant bits (simple fractions, for example, 0.25, 0.75), which result in shorter limit cycles (as above). Their solution for simple fractions was to offset the frequency by one LSB of their 24-bit accumulator. This resulted in a relative frequency change of less than 10^{-11}. We have performed similar experiments, using an offset of 2^{-16} rather than 2^{-24}, and we will now check those results with a constant sampling frequency and observe any change from previously observed spectrums.

With a first-order fractional-N, the region between the discrete spurs appears to be filled by noise, just as in Fig. 2.8. The fixed sampling rate did not make much difference in this simple case.

With a second-order modulator (fractional-N, ΣΔ), the spectrum (Fig. 2.30) is noise-like and is again similar to the spectrum with variable sampling in Fig. 2.13 except that no discrete spur shows (at ±625 kHz) and the noise peak is slightly lower. This is close to the theoretical shape. However, note that it is possible that discrete spurs would be visible if we reduced the noise bandwidth (see Section S.4).

The spectrum with a third-order modulator is noise-like, as shown in Fig. 2.31. This is close to theoretical, as we can see by comparing it with the theoretical curve in Fig. 2.22. Figure 2.32 shows results for a fourth-order fractional-N along with points from the theoretical curve. Note how the close-in PSD is reduced. The overall

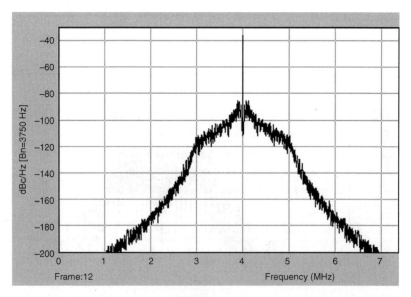

FIGURE 2.30 Output spectrum at 100.62515… with second-order fractional-N. Charge pump PD and integrator (CPPD&I) resampled at a constant rate. Compare with Fig. 2.13.

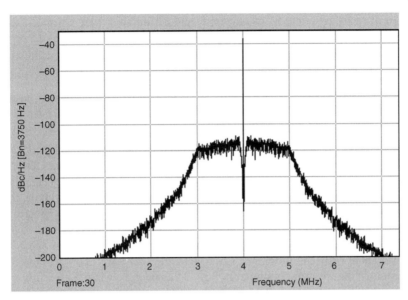

FIGURE 2.31   Output spectrum at 100.62515... with third-order fractional-N. This is close to the theoretical shape shown in Fig. 2.22. CPPD&I resampled at a constant rate, initial zero.

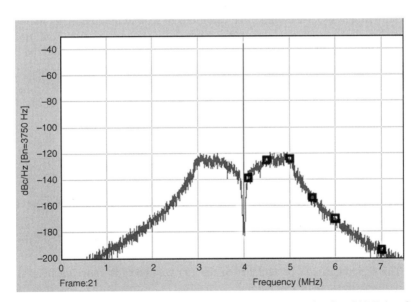

FIGURE 2.32   Output spectrum at 100.62515... with fourth-order fractional-N. Points from the theoretical PSD are also shown. CPPD&I resampled at a constant rate, initial zero.

spectrum, therefore, could be significantly better if the loop were narrower, so the noise at higher modulation frequencies could be more attenuated.

Therefore, with a small frequency offset from a simple fraction, the third- and fourth-order MASH circuits produce noise with a shape (versus offset frequency) that is close to theory, when the sampling rate is constant.

2.4.3 100.625 MHz with Seed

It has been observed [Filiol et al., 1998; Kozak and Kale, 2004] that initially setting the least significant bit (to 1) in the first accumulator eliminates discrete spurs for orders greater than 2 (although the latter indicates that the number of bits in the accumulators is also important, 14–22 bits being optimum to get theoretical results with the third-order MASH).[14] With an initial value (seed) of 2^{-16} in the first accumulator, the spectrum for the first-order fractional-N is again like Fig. 2.7. For a second-order circuit, the output is similar to that in Fig. 2.15, but the ±937.5 kHz spurs have dropped about 6 dB and those at ±1.875 MHz are 2 dB lower. For the third-order fractional-N, the spectrum is similar to Figs. 2.22 and 2.31, while for a fourth-order fractional-N it is close to Fig. 2.32. The latter two were with a small frequency offset. Thus, it appears that setting a small initial number in the first accumulator has an effect similar to that of changing the frequency by a small amount.

2.5 SUMMARY OF SPECTRUMS

The spectrums that we have compared are summarized in Table 2.3. Lines 11–19 show modulator orders 1–3, each with three variations of a small initial number in the first accumulator and a small frequency offset. Lines 1–8 are under the same conditions as lines 11–18 except that the former have varying sampling frequencies and the latter have fixed sampling frequencies and different PDs are employed.

2.6 SUMMARY

- Advantages of fractional-N synthesis include
  - reduced step size while maintaining wide loop bandwidth, and
  - reduced noise at the output because of the smaller divide ratio.
- Accumulators are used to convert the fractional divide number into an equivalent series of whole numbers.
- A simple fractional-N synthesizer needs a narrow bandwidth to filter the PD waveform caused by the sequencing of N-values.
- The bandwidth can be widened if the PD waveform is canceled by a signal derived from the accumulators.
- The canceling waveform can be obtained from the last stage of a MASH structure. Its effectiveness depends on how it is processed.

TABLE 2.3 Summary of Spectrums Compared

	Figure	Sample Rate	Modulator Order	$n_{\text{fract}} =$ 0.0625 +	Initial	Spur space (kHz)	Noise-Like Spectrum	PD Type
1	2.7	Variable	1	0	0	625	None	S&H
2	Like 2.7	Variable	1	0	2^{-16}	625	None	S&H
3	2.8	Variable	1	2^{-16}	0	625	Between	S&H
4	2.12	Variable	2	0	0	312.5	None	S&H
5	2.15	Variable	2	0	2^{-16}	312.5	Between	S&H
6	2.13	Variable	2	2^{-16}	0	±625	Yes	S&H
7	–	–	–	–	–	–	–	–
8	2.18	Variable	3	0	2^{-16}	±312.5	Yes	S&H
9	2.22	Fixed	3	0	2^{-16}	–	Yes	Reverse S&H
10	2.24a	Variable	3	0	2^{-16}	±312.5	Yes	CPPD&I
11	Like 2.7	Fixed	1	0	0	625	None	CPPD&I
12	Like 2.7	Fixed	1	0	2^{-16}	625	None	CPPD&I
13	Like 2.8	Fixed	1	2^{-16}	0	625	Between	CPPD&I
14	Like 2.12	Fixed	2	0	0	312.5	None	CPPD&I
15	Like 2.15	Fixed	2	0	2^{-16}	312.5	Between	CPPD&I
16	2.30	Fixed	2	2^{-16}	0	–	Yes	CPPD&I
17	2.27	Fixed	3	0	0	312.5	None	CPPD&I
18	Like 2.22	Fixed	3	0	2^{-16}	–	Yes	CPPD&I
19	2.31	Fixed	3	2^{-16}	0	–	Yes	CPPD&I
20	2.28	Fixed	4	0	0	156.25	None	CPPD&I
21	2.32	Fixed	4	2^{-16}	0	–	Yes	CPPD&I

- Fractions that require small-value bits to express have longer sequences, resulting in more closely spaced output spurs.
- Additional accumulators (i.e., higher order $\Sigma\Delta$ modulators) can be used to make the number sequences longer. This leads to finer spur spacing and they make the spectrum more noise-like.
- Third-order MASH structures with at least 7 bits tend to produce noise-like spectrums.
- MASH configurations lead to shaped quantization noise with low levels of low-frequency noise, more so for higher orders, making it easier to filter and eliminating the necessity to cancel the PD waveform.
- The PSD at the synthesizer output is shaped by MASH modulators as given by Eq. (2.3).
  - That equation assumes constant sampling rate, which does not occur without special provision.
  - Otherwise, the PSD is increased close to the carrier.
  - That increase is most noticeable with small $\overline{N}$, because the PSD from Eq. (2.3) increases faster with $\overline{N}$.

- Output PPSD is inversely proportional to f_{ref}.
- Simple fractions, without small-value bits, can produce widely spaced (not noise-like) spurs, even with long accumulators. This can be overcome by
  - using the next available number, which can result in only a small frequency change if the accumulators are long; or
  - initializing the first accumulator with a small number to increase the sequence length.
- The visibility of discrete spurs is a function of the noise bandwidth of the spectral display.

CHAPTER 3

OTHER SPURIOUS REDUCTION TECHNIQUES

We have considered two methods for reducing the discrete spurs that tend to occur when n_{fract} uses only more significant bits, causing a short repetition period in the accumulators. One was offsetting n_{fract} by a small number and the other was presetting a small odd number into the accumulator every time n_{fract} is changed[15] (a capability that some ICs may not provide). Here, we will consider two additional methods, randomly changing the least significant bit (LSB), at a rate of f_{ref}, and modifying the MASH structure to maximize the sequence length.

3.1 LSB DITHER

A popular method for eliminating discrete spurs is to dither (randomly change) the LSB at the input to the modulator. Figure 3.1 shows the results of dithering the LSB of a 10 bit modulator, using a pseudorandom sequence of 0s and 1s. The synthesizer output spectrum with such dithering is displayed above the spectrum obtained with 2^{-16} initially in the first accumulator. We see immediately that the center of the spectrum is shifted by the average value of the dither, half an LSB $(2^{-11} \times 10\,\text{MHz} = 4.88\,\text{kHz})$, and that the PSD near spectral center is increased significantly.

The value of $\mathcal{L}_\varphi$ with this dither is plotted as $q = 0$ curve in Fig. 3.2. Also shown is the quantization noise of a MASH-111 $\Sigma\Delta$ modulator ($p = 3$). These are the levels that appear in Fig. 3.1,[16] but less the loop response $H(\Delta f)$ (Fig. L.4), which is 0 dB at

Advanced Frequency Synthesis by Phase Lock, First Edition. William F. Egan.
© 2011 John Wiley & Sons, Inc. Published 2011 by John Wiley & Sons, Inc.

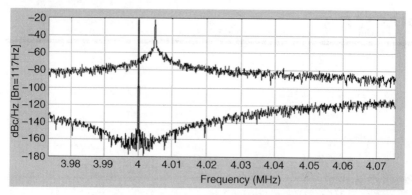

FIGURE 3.1   Output spectrum (L_φ) with dither in the LSB (2^{-10}) above a spectrum with no dither, but with 2^{-16} initial condition, third-order MASH-111, 100.625 MHz nominal.

spectral center but peaks near 0.06 MHz offset, increasing L_φ by 4 dB at that offset in Fig. 3.1.

While the unshaped dither noise is flat at the output of the modulator, the noise in the loop is proportional to a digitally integrated (accumulated) version of that noise (see Appendix Q). As with the quantization noise, this effectively divides the noise shape by a sinusoidal function of f_m, producing the -20 dB/decade slope that can be seen in Fig. 3.2. If we shape the dither in a filter that has the same transfer characteristic as a MASH modulator, the equation for dither noise will be similar to

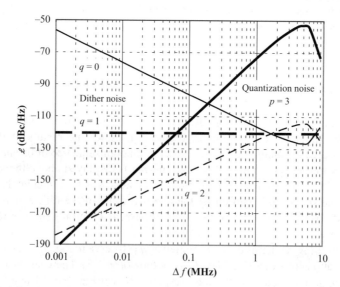

FIGURE 3.2   Single-sideband power spectral density (under small modulation index assumption) $\mathscr{L}_\varphi$ at synthesizer output from 2^{-10} dither and the quantization noise of third-order $\Sigma\Delta$. Dither has noise shaping of $(1 - z^{-1})q$. These curves do not include the attenuation due to the loop response that occurs as the higher values of Δf.

Eq. (2.3). The PPSD ($=2\mathcal{L}_\varphi$) for dither that has been shaped by multiplying the random sequence of 1s and 0s by $(1 - z^{-1})^q$ is

$$S_{\varphi d} = \frac{(2^{-n}2\pi \text{ rad})^2}{2f_{\text{ref}}} \left[2\sin\left(\pi\frac{f_m}{f_{\text{ref}}}\right) \right]^{2(q-1)} \tag{3.1}$$

(see Appendix M). We see that the dither noise decreases by 6 dB for each unit increase in the number of bits n. The value of q for simple, unshaped dither is 0, giving the dither its negative slope at low frequencies. Curves for higher order dither filters ($q>0$) are also shown in Fig. 3.2. The advantage of shaping the dither is apparent.

The simulated output spectrum is shown in Fig. 3.3 for $q=1$ and in Fig. 3.4 for $q=2$. The former provides a smooth noise plot at the level shown in Fig. 3.2, modified

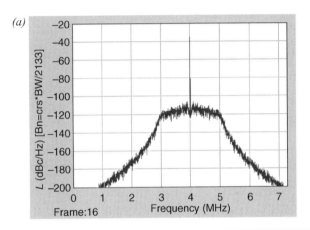

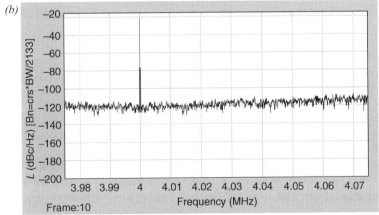

FIGURE 3.3 Output spectrum with shaped 2^{-10} dither, MASH-111 at 100.625 MHz, $q=1$. Noise bandwidth $B_n = 3750$ Hz in (a) and 117 Hz in (b).

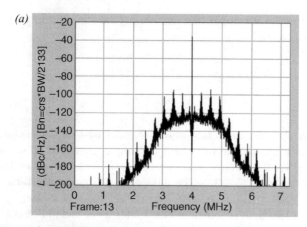

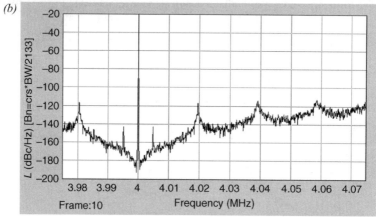

FIGURE 3.4 Output spectrum with shaped 2^{-10} dither, MASH-111 at 100.625 MHz, $q = 2$. Noise bandwidth $B_n = 3750$ Hz in (a) and 117 Hz in (b).

by $H(\Delta f)$, but the latter does not. Pamarti and Galton [2007] have found that in order for the dither to be effective in eliminating spurs,

$$q \leq p-2, \tag{3.2}$$

which implies that the $q = 2$ curve in Fig. 3.2 should not be used with the $p = 3$ MASH.

We could increase the order of the modulator to $p = 4$ and see the spectrum become smooth again.

Also apparent in Fig. 3.3b and Fig. 3.4b is the fact that, unlike unshaped dither, shaped dither does not shift the synthesized frequency. The shaping is digital differentiation, where each digit is followed by its negative, resulting in zero average value. Based on the results of simulations, it appears that this multiplication by a higher power of $(1-z^{-1})$ can also be accomplished by inserting the dither at the input to a later accumulator (e.g., A_2 rather than A_1 in Fig. 2.17). Another method for

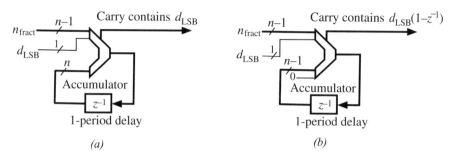

FIGURE 3.5   Inserting dither into first accumulator. Dither inserted in (b) is shaped, whereas that inserted in (a) is not.

shaping is shown in Fig. 3.5. In (a), a dithered LSB d_{LSB} is inserted into the first accumulator of a MASH along with more significant bits representing n_{fract}. In (b), the d_{LSB} is inserted into the register of the accumulator (not accumulated). Figure 3.6 shows how this produces the same effect as multiplying d_{LSB} by $(1 - z^{-1})$ at the accumulator's input. In both cases, we end up with d_{LSB} in the LSB of the register at r. Higher order shaping can similarly be accomplished with higher order first SQDSMs [Pamarti and Galton, 2007].

3.2   MAXIMUM SEQUENCE LENGTH

Hosseini and Kennedy [2007] indicate that a maximum sequence length can be obtained from a MASH-11... modulator, for all n_{fract} and initial conditions, by the proper choice of modulus implemented, as shown in Fig. 3.7. Here, the modulus is reduced by adding an integer a to the accumulator input after the carry is generated.

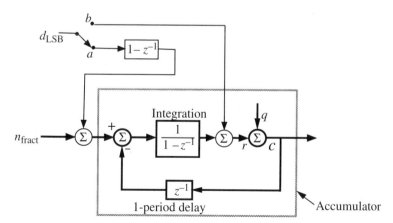

FIGURE 3.6   Illustration of the equivalence of inserting dither into the accumulator's register (b) to inserting it at the input (a) after multiplication by $(1 - z^{-1})$. The dither inserted into the register is not accumulated (not added to previous values).

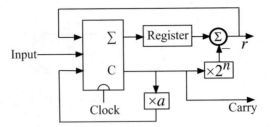

FIGURE 3.7   Accumulator in n-bit HK-MASH.

They used a value of $a = a_{HK}$ (Table X.1), the smallest integer that will reduce the modulus to a prime number. They obtain a resulting sequence length of

$$L_{sequ} = (2^n - a)^p, \tag{3.3}$$

where $n = n_a$ is the number of bits in the accumulators and p is the order of the $\Sigma\Delta$ modulator. The value of a_{HK} is smaller than 20 for $n \leq 24$, so the sequence length is

$$L_{sequ} \approx 2^{np} \tag{3.4}$$

for common values of n. For a third-order modulator, this is 2^{3n}, compared to a sequence length of 2^{n+1} with an odd seed (for 16 bits, 2.8×10^{14} versus 1.3×10^5).

When we use this technique, the basic step will be divided into fractions with a denominator that is not quite binary,

$$n_{fract} = \frac{N_{fract}}{2^n - a_{HK}} \tag{3.5}$$

(but binary fractions are not always desirable anyway).

Figure 3.7 shows how the accumulator diagrams in Fig. 2.1 are modified to implement the modulus reduction. To illustrate with a simple, if impractical, case, for a 4 bit accumulator with $n_{fract} = 1$ and $a = 1$, the value of r would repeat 0, 2, 3. These accumulators then fit into the multistage MASH structure as before (e.g., Fig. 2.17). We will call this architecture HK-MASH.

Figure 3.8 shows the spectrum from a synthesizer with an HK-MASH-11 modulator with 5 bit accumulators. Five bits were used to make the lengthening of the repetition sequence more observable. (Hosseini and Kennedy [2007] indicate that "Word-lengths less than 7 [bits] do a poor job of whitening the quantization noise" even in a MASH-111, but that is helpful for our observations.) For $n = 5$, the required a_{HK} equals 1, so

$$L_{seq} = (2^5 - 1)^2 = 31^2 = 961 \tag{3.6}$$

by Eq. (3.3) (by comparison, $16 \leq L_{seq} \leq 64$ for odd initial condition, by Section X.2). In Fig. 3.8a, we can see ripple peaks spaced by about 312.5 kHz,

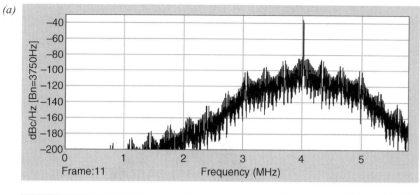

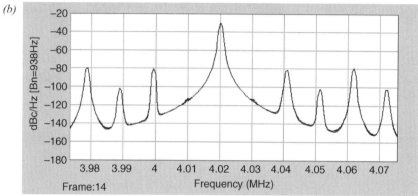

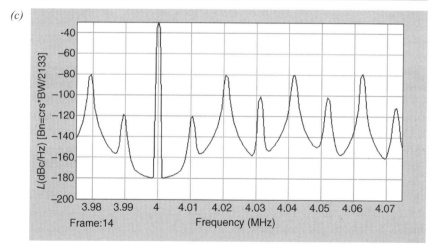

FIGURE 3.8   Output spectrum at 100.625 MHz (nominal) with modulo 31 accumulators in HK-MASH-11. CP&I resampled at a constant rate. Spectrum is expanded in (b) and recentered in (c).

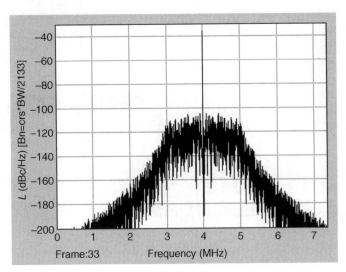

FIGURE 3.9   Synthesizer spectrum with 9 bit MASH-111 and LSB seed at 100.625 MHz. $B_n = 3750$ Hz.

as we have seen previously (e.g., Fig. 2.12). The expanded view in Fig. 3.8*b* shows spurs spaced by

$$\Delta f = 10\,\text{MHz}/961 = 10.40\ldots\text{kHz}. \tag{3.7}$$

If we had used 17 bits (for which a is also 1), the spacing would have been less than a millihertz, and if we had also used MASH-111, it would have been a few nanohertz.

Since 100.625 MHz has been converted down to 4 MHz in this display, we can also see [in (*b*)] that the central line is high in frequency by about 20 kHz. This is because each increment in n_{fract} now produces a frequency change that is higher by 32/31, relative to its value before the modulus was changed, an increase of 1/31, and 625 kHz/31 = 20.16 kHz. The step size has changed from 10 MHz/32 = 312.5 kHz to 10 MHz/31 = 328.58...kHz. This does not represent an error, rather just a difference in the relationship between f_{fract} and n_{fract}. Of course, with more bits, the shift would be less apparent.

In Fig. 3.8*c*, the spectrum has been shifted so that the main line is centered in a bin, reducing the broadening artifact shown in (*b*) (see Section S.3). Other lines that are not centered in bins are still broadened.

Figure 3.9 shows the output from our synthesizer with MASH-111 using 9 bit accumulators and an LSB seed. Figure 3.10 shows the output from a 9 bit HK-MASH-111 with no seed. The improvement is obvious. Comparing the spectrums of modulator outputs (not synthesizer outputs), Hoseini and Kennedy indicate that "a 9 bit HK-MASH is as effective as an 18 bit seeded MASH." The output from a 16 bit MASH-111 with LSB seed in Fig. 3.11 does appear similar to Fig. 3.10.[17] Thus, the

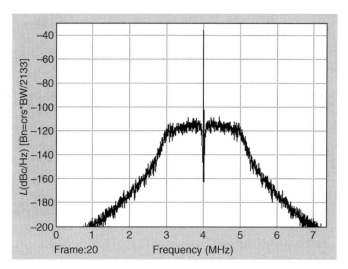

FIGURE 3.10   Synthesizer spectrum with 9 bit HK-MASH-111, no seed, at 100.6287... MHz. $B_n = 3750$ Hz.

HK-MASH configuration has potential for a significant reduction in the number of circuits and, therefore, also power consumption.

The circuit modification for $a = 1$ is particularly simple. The carryout is delayed by 1 clock and inserted into the carry-in of the adder in the accumulator. For larger values of a, the adder would be modified to take a third input. There are five values of n between 5 and 19 that use $a_{HK} = 1$ (Section X.1).

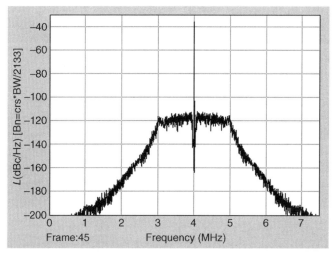

FIGURE 3.11   Synthesizer spectrum with 16 bit MASH-111 and LSB seed at 100.625 MHz. $B_n = 3750$ Hz.

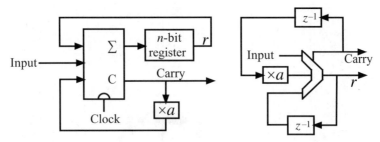

FIGURE 3.12 Two representations of a simplified HK-MASH modulator that is useful under restricted circumstances.

With $a = 1$, Fig. 3.7 can be further simplified to Fig. 3.12, where the subtraction of 2^n is accomplished by the normal rollover of the n-bit register. For this substitution to be accurate, the resulting modulo 2^n sum must be the same as subtraction of 2^n from the sum, meaning that the sum must never equal or exceed 2^{n+1}. (2^{n+1} modulo $2^n = 0$, but $2^{n+1} - 2^n = 2^n$.) Even with the maximum $2^n - 1$ in an accumulator register, a maximum input of $2^n - 1$, and the addition of $a = 1$, the resulting sum meets this criterion,[18] so the configuration in Fig. 3.12 can be used for all HK-MASH stages when $a = 1$. Moreover, for any value of a, the first stage can use Fig. 3.12, because the maximum input to the first stage is $2^n - 1 - a$, so the maximum sum, even with a added, does not reach 2^{n+1}.

However, for larger values of a, should an accumulator register contain the maximum $2^n - 1$ at the same time that a carry has been produced, and with the input from the previous accumulator also being maximum, the next sum (including a) would equal or exceed 2^{n+1}. We have not shown that this condition is possible for all $a_{HK} > 1$ and all accumulators, but it seems very likely for large values of a. It may still be practical to use Fig. 3.12 for smaller values of a if the inaccuracy seldom occurs, since only the first accumulator affects the average output of the MASH circuit. Extensive simulation to verify that the noise spectrum be acceptable would then be warranted (`HandKsimple.mdl` can be used for this).

3.3 SHORTENED ACCUMULATORS AND LOWER PRIMES

While the literature tends to concentrate on accumulators of binary lengths, many synthesizer ICs permit the designer to choose the denominator in n_{fract} practically arbitrarily [Banerjee, 2006]. However, selecting a prime number modulus does not produce the same results as obtained with the HK-MASH configuration, as we can see by comparing Fig. 3.13 with Fig. 3.10. The modulus was 509 in both cases, but while the inputs to accumulators after the first cover the range of 0–508 in MASH, they extend from 0 to 511 in the HK-MASH.

We have, so far, seen no obvious reason why the HK-MASH structure cannot be used to achieve prime moduli, even using values of a that are not the smallest that will produce a prime (i.e., a_{HK}), except for the complexity of implementing Fig. 3.7.

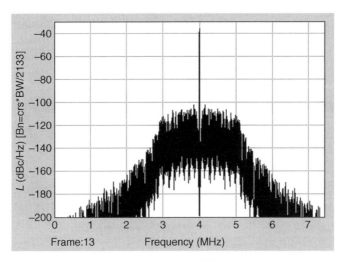

FIGURE 3.13   Synthesizer spectrum with modulus 509 MASH-111, odd seed, at 100.6287...
MHz; $n_{fract} = 32/509 = 0.06287...$.

Large values of a will produce more instants where Fig. 3.12 is not equivalent to
Fig. 3.7, so the spectral changes can be expected to be more severe if the simple
circuit is used with large a. Simulations were performed to compare the results of the
configurations of Figs. 3.7 and 3.12 (using HandK.mdl and HandKsimple.
mdl, respectively). The accumulators had 9 bits (512 capacity) and the noise
bandwidth B_n was 3750 Hz, as in Figs. 3.9–3.11 (but the observed spectrums were
expanded to cover just 3.5–5.5 MHz). Inputs of $N_{fract} = 1$, 32, and 256 were
employed ($n_{fract} = N_{fract}/512$). The greatest differences were observed with N_{fract}
$= 1$. Values of a corresponding to various prime moduli M were employed. With
$a = a_{HK} = 3$, the spectrums obtained with the two configurations appeared identical,
but there was a small noticeable difference with $a = 9$ and $N_{fract} = 1$. At $a = 21$
($N_{fract} = 1$), the simplified circuit showed spurs at offsets that were multiples of
10 MHz/M. Surprisingly, with $a = 25$, 33, and 45, *both* configurations showed such
spurs. It appears that a small value of a is important, even when the circuit in Fig. 3.7
is used, and Fig. 3.14 offers a clue as to why this is so.

The tuning voltage in the top window of Fig. 3.14 produces the spurs shown in
Fig. 3.15. The time between peaks in Fig. 3.14 is 46.7 µs, the reciprocal of the spur
spacing (10 MHz/M) seen in Fig. 3.15. The second window in Fig. 3.14 shows the
output from the HK-MASH-111 modulator and the windows below that show the
contents of the three registers. We can see the first register (middle window) advancing
linearly in response to the unity input. We can also observe how, on the next clock after
its contents go to 0, the level jumps, as $a = 45$ is added. This register provides the input
to the second register, and something unusual can be seen there, when the first register
is near maximum. The contents of the second register climb as high as 2.5. This is a
consequence of the process that we discussed above, when describing why the
simplified circuit acts differently from the proper HK circuit. While the proper circuit,

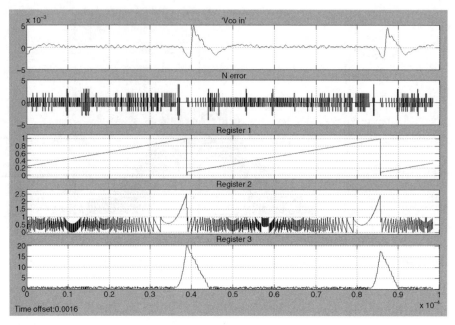

FIGURE 3.14   Time waveforms for HK-MASH-111, 9 bits, $a = 45$, $N_{\text{fract}} = 1$. *Top to Bottom*: Tuning voltage; MASH output; first, second, and third accumulators' contents.

which we are here observing, performs correctly, the combination of high-value inputs from the first accumulator and the relatively large value of a cause the contents of the second accumulator to reach a higher than usual value. This value then provides input to the third accumulator, which reaches even higher peaks; a value of 20 can be observed in the displayed segment of the time plot. These large values, occurring in

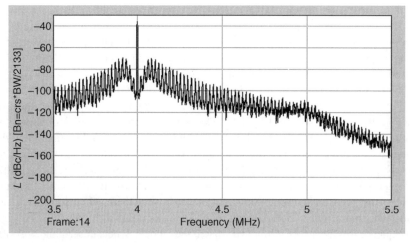

FIGURE 3.15   Spectrum for HK-MASH-111, 9 bits, $a = 45$, $N_{\text{fract}} = 1$. $B_n = 3750$ Hz.

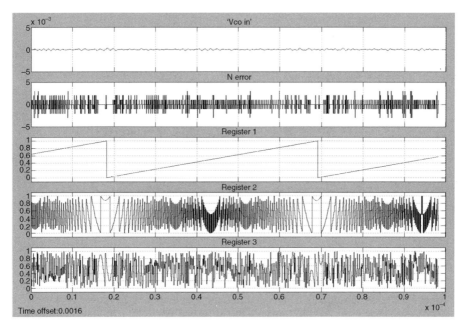

FIGURE 3.16 Time waveforms for HK-MASH-111, 9 bits, $a = 3$, $N_{\text{fract}} = 1$.

synchronism with the peaks in the first accumulator, produce transients in the tuning voltage that result in the observed spurs.

This does not mean that the overall sequence is not long—we can see that the peaks on the tuning voltage are not really identical—but just that a process is occurring that produces spurs synchronized to the first accumulator, in spite of the length of the overall sequence. Figure 3.16 shows a similar time sequence with $a = a_{\text{HK}}$, the minimum value that produces a prime modulus. The three upper windows have the same scale as in Fig. 3.14 and as such we can see that the low-frequency modulation is gone, but the lower two are expanded and so we can see that the accumulator contents now stay within the usual range of 0–1.

It is apparent that we must exercise caution in using larger values of a, even with the full, proper, HK-MASH configuration.

3.4 LONG SEQUENCE

Song and Park [2010] have proposed another architecture for obtaining long sequences, one in which the carry from one MASH stage causes an LSB to be added to the next stage, apart from, and in the same manner as, the addition of its register contents (Fig. 3.17). This produces a maximum length sequence in the accumulator to which the LSB is added. As a result, the length of the overall sequence is

$$L_{\text{sequ}} = N_1 N_a^{p-1}, \tag{3.8}$$

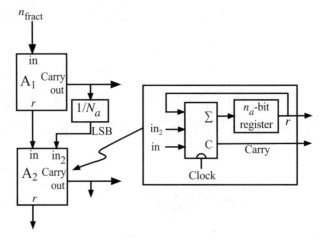

FIGURE 3.17   SP-MASH concept.

where $N_1 \geq 2$ is the length of the sequence in the first accumulator and $N_a = 2^{n_a}$ is the capacity of each subsequent n_a-bit accumulator. The length of the sequence in the first accumulator, N_1, depends on the value of n_{fract}. We will call this architecture SP-MASH. Compare Eq. (3.8) with Eq. (3.4) for the HK-MASH.

A second difference between the two architectures is that in SP-MASH,

$$n_{\text{fract}} = N_{\text{fract}}/2^{n_a}, \tag{3.9}$$

as is the case for ordinary MASH modulators, whereas n_{fract} is given by Eq. (3.5) for HK-MASH.

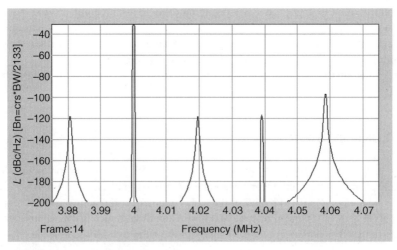

FIGURE 3.18   Synthesizer spectrum for 5 bit SP-MASH-11, 100.625 MHz. CP&I resampled at a constant rate.

The first accumulator can be shorter than the others—it only needs to be long enough to accommodate the input fractions being used—in which case N_a in Fig. 3.17 refers to the others and r from the first accumulator would be added to the MSBs of the second accumulator.

An SP-MASH-111 synthesizer with 9 bit accumulators ($n_{\text{fract}} = 0.0625 + 2^{-9}$) has about the same sequence length as the 9 bit HK-MASH-111, whose spectrum is shown in Fig. 3.10, and its spectrum appears similar. However, so does the spectrum of the SP-MASH-111 when n_{fract} changes to 0.0625, producing a shorter sequence. The difference is not apparent with a display of this B_n and magnification. If we compare Fig. 3.18 with Fig. 3.8c, we can see the difference. These are spectrums of 5 bit SP-MASH-11 and HK-MASH-11 synthesizers, respectively, with $n_{\text{fract}} = 0.0625 = 2^{-4}$ and Bn $\approx$ 469 Hz. The sequence length for the former is $2^4 \times 2^5 = 512$, whereas it was 961 for the latter [Eq. (3.4)]. This leads to a spur spacing of 10 MHz/512 = 19.53... kHz in Fig. 3.18 compared to 10.40... kHz [Eq. (3.5)] in Fig. 3.8c.

The amount of circuitry required for HK-MASH is greater than what is required for SP-MASH when each uses the same size accumulators. However, since the sequence length is shorter with SP-MASH when n_{fract} is short (lacks small-value bits), it might require longer accumulators, so the trade-off is not so simple.

3.5 SUMMARY

- Spurs can be attenuated by pseudorandom dither in the LSB of N_{fract}:
  - The dither produces additional noise proportional to the value of the LSB.
  - It can be shaped similar to quantization noise.
  - There is an optimum order for effective shaping.
- The length of a $\Sigma\Delta$ sequence can be maximized by using the HK-MASH structure in which the moduli are prime numbers.
- Nonbinary moduli can be easily realized in some ICs.
- Using a prime modulus does not afford the benefits of the HK-MASH unless the HK modulator structure is employed.
- Spurious suppression can be lost if HK-MASH moduli deviate too greatly from binary values.
- The SP-MASH architecture also affords long sequences and uses binary moduli.

CHAPTER 4

DEFECTS IN $\Sigma\Delta$ SYNTHESIZERS

In this chapter, we consider the effects of some of the imperfections that can occur in the implementation of a $\Sigma\Delta$ frequency synthesizer. To do this, we will first create theoretical noise models, adding quantization noise to sources such as those that we studied in Sections F.3.8 and F.3.9 and combining them all in several synthesizer loops in order to study how they affect the output PPSD. This is similar to what we did in Section F.3.10, except for the addition of quantization noise. We will also use noise levels here that are appropriate for available ICs. Then, we will consider how additional noise sources that are related to $\Sigma\Delta$ synthesis affect the closed loop performance of our models.

4.1 NOISE MODELS

4.1.1 VCO Noise

Figure O.1 (Appendix O) is a plot of oscillator spectrums. It joins the plots of Fig. F.3.38*a–e*. Recall that the plots are normalized in the way they are because they tend to fall -20 dB/decade to a noise floor with the break at

$$f_m = \frac{f_{osc}}{2Q}, \tag{4.1}$$

Advanced Frequency Synthesis by Phase Lock, First Edition. William F. Egan.
© 2011 John Wiley & Sons, Inc. Published 2011 by John Wiley & Sons, Inc.

where f_{osc} is the oscillator's frequency and Q is the loaded Q. Therefore, if we plot against f_m/f_{osc}, oscillators with similar noise floors and Q's will have similar S_φ in their -20 dB/decade regions, regardless of the value of f_{osc}. This helps us scale the curves for different frequencies.

Curves $1a$ and 10 bracket most of the noise curves in Fig. F.3.38a–e and form "points-of-reference" between plots. The lower noise curves in Fig. O.1 are for miniature oscillators using planer resonators. The others are LC oscillators in ICs. Other oscillators sometimes used in ICs are ring oscillators and astable multivibrators. Their PPSDs are shown in Fig. F.3.38e; they are significantly noisier (up to 30 dB) even than the $1a$ reference curve.

In the -20 dB/decade region, S_φ is proportional to f_{osc} at a given f_m, but the noise floor and the frequency at which the slope changes to -30 dB/decade do not so scale.

VCO noise appears at the loop output after multiplication by the error response,

$$S_{out,vco} = S_{vco,\varphi}|1-H(f_m)|^2, \tag{4.2}$$

where $S_{vco,\varphi}$ is the PPSD of the VCO (open loop) and $NH(f_m)$ is the closed-loop forward transfer function of the loop.

4.1.2 Basic Reference Noise

By the basic reference we mean the oscillator at f'_{ref} from which the loop reference at f_{ref} is derived (see Fig. F.1.19c). The PPSD that appears at the output due to the basic reference is

$$S_{out,ref'} = S_{ref,\varphi}|NH(f_m)|^2 = S_{ref',\varphi}\left(\frac{N}{M}\right)^2|H(f_m)|^2 \tag{4.3}$$

$$= S_{ref',\varphi}\left(\frac{F_{out}}{F'_{ref}}\right)^2|H(f_m)|^2, \tag{4.4}$$

where $S_{ref',\varphi}$ is the PPSD of the basic reference and M is the divider ratio of the divider that follows the basic reference and outputs f_{ref}.

4.1.3 Equivalent Input Noise

Noise that appears at the input to the loop, or its equivalent, is multiplied by the closed-loop forward gain. This may be noise that appears elsewhere in the loop after being multiplied by the gain from its point of appearance to the input (see Appendix F.6.A).

The output PPSD due to component noise is

$$S_{out,in} = S_{in,\varphi}|NH(f_m)|^2, \tag{4.5}$$

where $S_{in,\varphi}$ represents the equivalent PPSD at the loop input. This, and other PPSDs that contain $|NH(f_m)|^2$, are proportional to N^2, which is the low-frequency power gain.

While it is important, when designing $\Sigma\Delta$ synthesizer circuits, to understand the theory of potential noise sources, characterization of existing synthesizer ICs is of great value when using them to meet system requirements. Fortunately, Banerjee [2006] has provided detailed noise information on National Semiconductor synthesizer ICs. This noise, referred to the loop input, can be characterized as[19]

$$\mathcal{L}_{\text{ref,bj}} = \left[k_1 f_{\text{ref}} + k_{11} \frac{f_{\text{ref}}^2}{f_m} \right] \left[1 + \frac{I_{\text{knee}}}{I_{\text{cp}}} \right] \text{Hz}^{-1}. \tag{4.6}$$

Parameters k_1 and k_{11} are the relative densities of white and flicker noises, respectively, at 1 Hz f_{ref} and (for k_{11}) 1 Hz f_m. If we multiply Eq. (4.6) by $(f_{\text{out}}/f_{\text{ref}})^2$ to obtain the resulting noise level at the loop output (at low f_m), we find that the white noise level at the output is reduced when f_{ref} increases. Note that this noise does not include effects due to $\Sigma\Delta$ modulation. Later we will consider the sources for this observed noise as well as the effects of $\Sigma\Delta$ modulation on it.

4.1.4 $\Sigma\Delta$ Quantization Noise

PPSD at the synthesizer output due to $\Sigma\Delta$ quantization from a pth-order MASH modulator is, from Eqs. (2.2) and (2.3),

$$S_{\text{out},\Sigma\Delta} = \frac{(2\pi \text{ rad})^2}{6 f_{\text{ref}}} \left[2 \sin\left(\pi \frac{f_m}{f_{\text{ref}}} \right) \right]^{2(p-1)} |H(f_m)|^2. \tag{4.7}$$

Since the basic phase quantum is one cycle at f_{out}, $S_{\text{out},\Sigma\Delta}$ is independent of f_{out}, for which reason Eq. (4.7) does not depend on N.

4.1.5 Parameter Dependence

The parameter dependence of these noise sources, as they affect the synthesizer output PPSD, is shown in the first six rows of Table 4.1. (We will consider the remaining rows later). We show dependence on f_{out} and f_{ref} rather than on N; any two of these three variables could have been used.

4.1.6 Synthesizer Output Noise

Let us now create a "typical" model for a frequency synthesizer that uses a MASH-111 modulator and is suitable for integration. This will allow us to see how the various sources of noise interact to produce an overall noise profile. There would be criteria present in a system design that would place limitations on the noise profile and guide our trade-offs, but in their absence, we will attempt to adjust the loop parameters to minimize the noise in some sense and hope to gain a better understanding of the interplay of the parameters and noise sources.

TABLE 4.1 Parameter Dependence of Noise Sources at the Output

		$\|H\|$ factor	f_{out}	f_{ref}	f_m	
$\mathcal{L}_{out,vco}$ Section 4.1.1	VCO noise	$\|1-H(f_m)\|^2$	f_{out}^2 [a]			
$\mathcal{L}_{out,ref'}$ Section 4.1.2	Reference (basic) noise	$\|H(f_m)\|^2$	f_{out}^2			$(f_{ref}')^{-2}$
$\mathcal{L}_{out,in}$ Eq. (4.5)	Component noise [b]	$\|H(f_m)\|^2$	f_{out}^2	f_{ref}^{-2}		
$\mathcal{L}_{out,bw}$ Eq. (4.6)	Banerjee's white noise [c,d]	$\|H(f_m)\|^2$	f_{out}^2	f_{ref}^{-1}		
$\mathcal{L}_{out,bf}$ Eq. (4.6)	Banerjee's flicker noise [c]	$\|H(f_m)\|^2$	f_{out}^2		f_m^{-1}	
$\mathcal{L}_{out,\Sigma\Delta}$ Eq. (4.7)	ΣΔ Quantization noise [e]	$\|H(f_m)\|^2$		f_{ref}^{1-2p}	$f_m^{2(p-1)}$	
$\mathcal{L}_{out_{excess}}$ Eq. (2.4)	Variable sampling rate [e]	$\|H(f_m)\|^2$	f_{out}^{-2}	f_{ref}^{-1}	f_m^2	
$\mathcal{L}_{out,d}$ Eq. (3.1)	LSB dither [e]	$\|H(f_m)\|^2$		f_{ref}^{1-2q}	$f_m^{2(q-1)}$	
$\mathcal{L}_{out,u3}$ Section 4.2.3	Mismatch of CP currents	$\|H(f_m)\|^2$		f_{ref}^{-1}		

[a] At small enough f_m with fixed noise floor and loaded Q.
[b] Referenced to the loop input.
[c] Also depends on CP current level and includes sampled jitter [Eq. (4.13)], charge pump current noise [Eq. (4.16)].
[d] Also depends on CP current level and includes timing imperfections (Section 4.3.2).
[e] At $f_m \ll f_{ref}$.

For the basic reference oscillator, we will use the 5 MHz crystal standard of curve 6 in Fig. F.3.38a, which has a noise floor of $S_\varphi = -149$ dBr/Hz. If it should become important in a design, there are ways to get a better noise floor (e.g., *PLB*, Section 12.6), but this will serve to show how the reference noise influences the synthesizer output.

We will use Banerjee's noise model, calling this noise density $\mathcal{L}_{bj}$, with parameter values that are neither the best nor the worst available. In his table of phase noise parameters for National Semiconductor PLLs [Banerjee, 2006, Table 14.1, p. 105], 10^{-22}/Hz $\leq k_1 \leq 4.7 \times 10^{-21}$/Hz and 10^{-25}/Hz $\leq k_{11} \leq 2.5 \times 10^{-23}$/Hz, so we will use

$$k_1 = 2 \times 10^{-21}/\text{Hz} \quad \text{and} \quad k_{11} = 10^{-23}/\text{Hz},$$

without the noise plateau below 1 kHz that he has observed in some units.

We will use the quantization noise for MASH-111 from Eq. (2.3), $p = 3$.

We will use VCO noise from the 799 MHz LC IC oscillator of curve 34 in Fig. O.1 in the -6 dB/octave region with the noise floor from another LC IC oscillator, curve 33, since curve 34 does not extend to its floor. We will use the noise floor from curve 33 for all VCOs and set the frequency at which the slope changes from -30 dB/decade to -20 dB/decade at 1 kHz, just slightly lower[20] than that in curve 34, but will otherwise scale the PPSD with frequency.

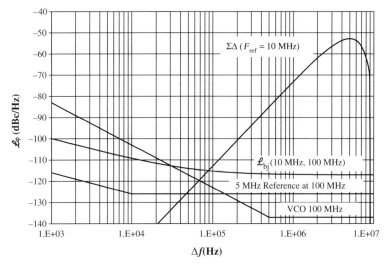

FIGURE 4.1 Noise sources for $F_{ref} = 10$ MHz and $F_{out} = 100$ MHz. These are the levels at f_{out} without attenuation due to the loop response.

4.1.6.1 Nominal Parameters

The four noise sources adjusted for $F_{out} = 100$ MHz and $F_{ref} = 10$ MHz are shown in Fig. 4.1. The two oscillator noise curves are tangential approximations. The VCO noise will be multiplied by $|1 - H|^2$, attenuating it at lower frequencies, but the other three noise components will be multiplied by $|H|^2$, attenuating them at high frequencies.

The response curves for our current loop model, which are given in Section L.1, show that it has very small gain, and phase, margins. While this is not a problem for our experiments, we would want more margins in a practical circuit. Moreover, the tangential (straight line) approximations that we would like to use in our analysis do not fit very well when the margins are so low. Therefore, we will double the pole frequencies and halve the zero frequency, while maintaining the 58 kHz unity gain bandwidth of our model, in order to increase the margins. This variation is described in Section L.2 and the straight line approximations for the original and the more stable configuration are shown in Fig. 4.2.

Using these loop parameters and the noise sources of Fig. 4.1, we obtain the various PSDs at the synthesizer's output, as shown in Fig. 4.3.

We see that the $\Sigma\Delta$ quantization noise protrudes above the VCO noise, so we will decrease the loop bandwidth (and decrease the filter frequencies by the same factor in order to maintain margins) to bring it down to the level of the VCO noise, which occurs at a unity gain bandwidth of $f_L = 13,340$ Hz. This results in the noise levels in Fig. 4.4.

This may look better, but the truly optimum solution depends on the system requirements. We may also want to search for an IC with a lower level of $\mathcal{L}_{bj}$. The minimum noise profile, in the sense in which we considered it in Section F.3.10, would occur with an f_L of about 30 kHz, where the VCO noise crosses $\mathcal{L}_{bj}$ in Fig. 4.1. But this just becomes an optimization between the VCO noise and noise effectively injected at

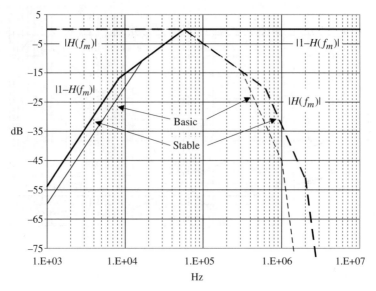

FIGURE 4.2   Basic and more stable error and forward gains: tangential approximations.

the reference. It would not involve the ΣΔ quantization noise at all. Here we are seeking a wide bandwidth, limited by the objective of keeping the quantization noise from making a significant contribution.

4.1.6.2 Higher F_{out}   ΣΔ Synthesizers are often used at several gigahertz; we will raise the frequency to 1 GHz to observe the effect of the decade increase in output

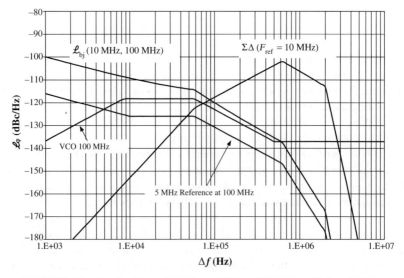

FIGURE 4.3 $\mathcal{L}_{out}$ for $F_{ref} = 10$ MHz and $F_{out} = 100$ MHz with $f_L = 58$ kHz.

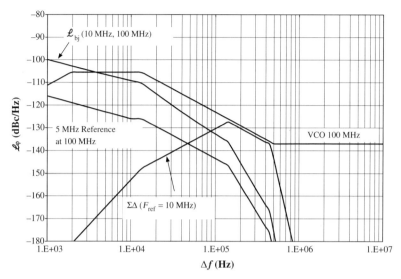

FIGURE 4.4 $\mathcal{L}_{\text{out}}$ for $F_{\text{ref}} = 10\,\text{MHz}$ and $F_{\text{out}} = 100\,\text{MHz}$ with $f_L = 13.34\,\text{kHz}$.

frequency. Figure 4.5 shows the noise levels at 1 GHz and Fig. 4.6 shows them at the output of the loop with $f_L = 29\,\text{kHz}$, which is half of our beginning bandwidth (Fig. 4.2) but about twice what we used with 100 MHz output frequency (Fig. 4.4). Again we have chosen a bandwidth that causes the quantization noise to just touch the VCO noise.

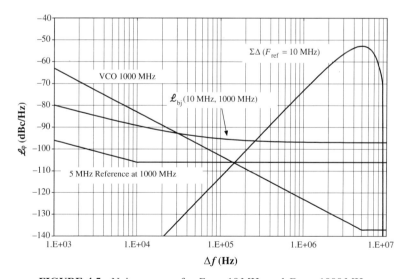

FIGURE 4.5   Noise sources for $F_{\text{ref}} = 10\,\text{MHz}$ and $F_{\text{out}} = 1000\,\text{MHz}$.

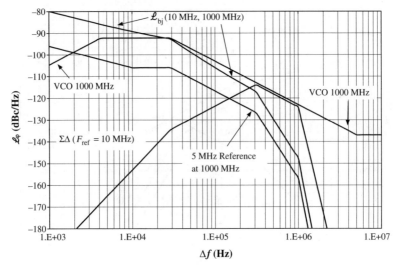

FIGURE 4.6 $\mathcal{L}_{\text{out}}$ for $F_{\text{ref}} = 10$ MHz and $F_{\text{out}} = 1000$ MHz with $f_L = 29$ kHz.

4.1.6.3 Higher F_{ref}   Increasing F_{ref} by 10 times will improve $\mathcal{L}_{\text{bj}}$ some and improve $S_{\text{out},\Sigma\Delta}$ considerably, because not only will its peak value be reduced but also the peak will be shifted to a higher frequency, as shown in Fig. 4.7. The output PSD is shown in Fig. 4.8 with the nominal (original) 58 kHz f_L and in Fig. 4.9 with a higher loop bandwidth, one that barely keeps the quantization noise below other noise.

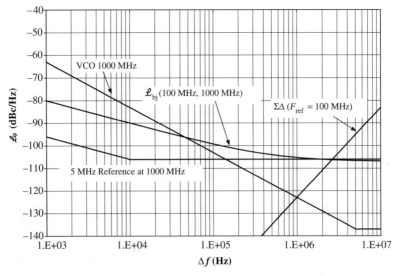

FIGURE 4.7   Noise sources for $F_{\text{ref}} = 100$ MHz and $F_{\text{out}} = 1000$ MHz. The peak of the quantization noise has shifted off the graph to 5×10^7 Hz.

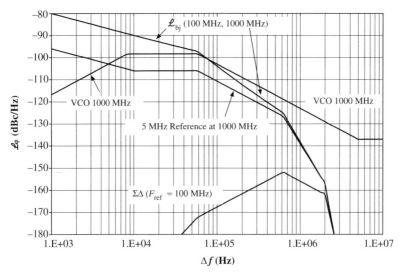

FIGURE 4.8 $\mathcal{L}_{out}$ for $F_{ref} = 100\,\mathrm{MHz}$ and $F_{out} = 1000\,\mathrm{MHz}$ with $f_L = 58\,\mathrm{kHz}$.

4.1.6.4   Summary   Figure 4.10 shows the maximum noise levels in the four output spectrums described above and whose properties are given in Table 4.2. These *maximums* are close to the *total* noise levels except where several noise sources have about the same level (see Figs. 4.4, 4.6, 4.8, and 4.9). Two equal sources increase the total by 3 dB and three equal sources produce a 5 dB increase relative to the values shown. We are using maximum levels rather than totals for simplicity. This should have little effect on the indicated rms phase noise values,[21] since most of the phase noise comes from the lower frequencies where one source usually dominates. Both the

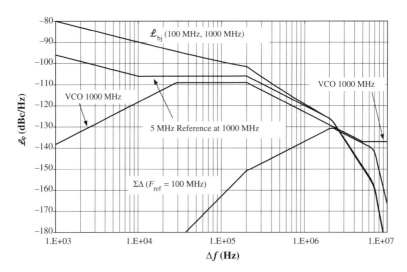

FIGURE 4.9 $\mathcal{L}_{out}$ for $F_{ref} = 100\,\mathrm{MHz}$ and $F_{out} = 1000\,\mathrm{MHz}$ with $f_L = 200\,\mathrm{kHz}$.

TABLE 4.2 Properties of Spectrums

Curve	F_{ref} (MHz)	F_{out} (MHz)	$f_L{}^a$ (kHz)	Integrated Phase[b] (Degree)
1	10	100	13.3	0.08
2	10	1000	29	0.62
3	100	1000	58	0.57
4	100	1000	200	0.72

[a]Frequency at unity open-loop gain.
[b]Phase from integrating PPSD over the frequency range shown.

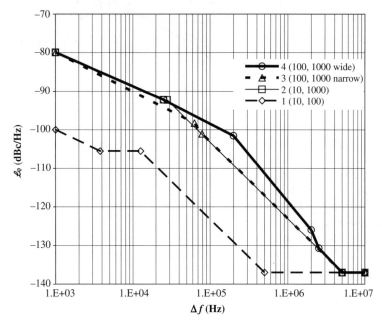

FIGURE 4.10   Summary of spectrums. Maximum levels shown. Parameters are F_{ref}, F_{out} in MHz.

quantization noise density and the white (higher frequency) component of $\mathcal{L}_{\varphi,bj}$ decreased with the increase in F_{ref}, but we took advantage of this to widen the bandwidth.

As Δf decreases below 1 kHz (the lowest value shown in Fig. 4.10), the spectrums climb at 10 dB/decade, due to the flicker noise in $\mathcal{L}_{bw}$, until about 0.16 Hz, below which the −30 dB/decade noise of the basic reference dominates.

4.2 LEVELS OF OTHER NOISE IN ΣΔ SYNTHESIZERS

Now that we have some models to show the synthesizer's output noise, let us see how the level of noise produced by some other quantifiable noise sources affects these outputs.

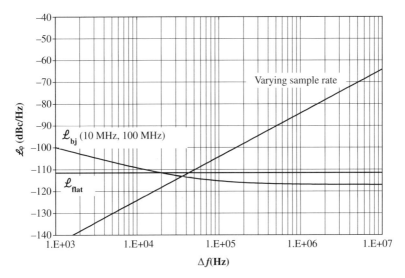

FIGURE 4.11 $\mathcal{L}_{bj}$ from Fig. 4.1 plus defect noise.

We begin by looking again at the 10, 100 ($= F_{ref}$, F_{out} in MHz) case in Figs. 4.1 and 4.4. For simplicity, we delete all the noise sources that will be multiplied by $H(f_m)$ except the dominant (at low frequencies) $\mathcal{L}_{bj}$. Then, we add a horizontal line to represent the level $\mathcal{L}_{flat}$ of a flat noise contribution from as yet undetermined sources. See Fig. 4.11 (ignore the third line for now). We set the flat noise 6 dB below the maximum VCO level in Fig. 4.4 so that it will increase the total noise in that region [where $H(f_m) \approx 1$] by only 1 dB. We will determine the magnitude of two effects that would produce this flat noise level.

4.2.1 Dither

The dither that we will use produces noise given by Eq. (3.1) with $q = 1$, the maximum for effective dither, according to the restriction of Eq. (3.2). As a result, the noise due to the dither is flat and will produce the level $\mathcal{L}_{flat}$ when the dithered LSB has value $2^{-9.04}$. Of course, there is no such bit, so we would have to use the tenth, or less, significant bit to ensure obtaining a lower noise level.

4.2.2 Varying Sample Rate

Next we add noise representing the effects of a varying sample rate, according to Eq. (2.4). This is the sloped line in Fig. 4.11 and it results in closed-loop output noise that exceeds the maximum noise in Fig. 4.4, as we can see in Fig. 4.12. (The noise sources that we removed in drawing Fig. 4.11 are also dropped here.) We see that the response to $\mathcal{L}_{flat}$ is everywhere below the maximum noise level by at least 6 dB, but the excess noise due to varying sampling rate exceeds the otherwise maximum level of $\mathcal{L}_{out}$ in a small region. Again, this excess noise occurs in the presence of ΣΔ modulation, and is not included in $\mathcal{L}_{bj}$.

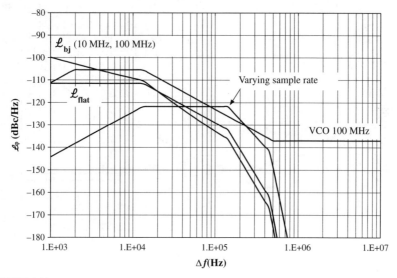

FIGURE 4.12 $\mathcal{L}_{\text{out}}$ for $F_{\text{ref}} = 10$ MHz and $F_{\text{out}} = 100$ MHz, $f_L = 13.34$ kHz, with added noise due to defects.

4.2.3 Mismatched (Unbalanced) Charge Pumps

Here we will show that mismatch in the PFD charge pumps is likely to produce a noise floor that exceeds $\mathcal{L}_{\text{flat}}$ in Fig. 4.11. Figure 4.13 shows results of simulations of MASH synthesizers where the positive charge pump current was $(1 + k_u)$ greater than the negative charge pump current.

The measured quantization noise tracks Eq. (2.3) when the currents are matched, but a 1% mismatch creates a noise floor, one that is slightly higher for a fourth-order modulator than for a third-order one. Also plotted is the noise when the PD output is not resampled (i.e., variable sampling rate) and the theoretical noise due to variable sampling rate (Section 2.3.3), so we see that the two effects add or are superimposed. These noise floors, obtained through simulation, are close to theoretical levels, given in Appendix U, for $f_m \ll f_{\text{ref}}$, as

$$\mathcal{L}_{\text{out},u3}(k_u) \approx \frac{k_u^2}{f_{\text{ref}}}(1 + n_{\text{fract}})2.5 \qquad (4.8)$$

for MASH-111 and

$$\mathcal{L}_{\text{out},u4}(k_u) \approx \frac{k_u^2}{f_{\text{ref}}}(1 + 0.5n_{\text{fract}})6.25 \qquad (4.9)$$

for MASH-1111. In Fig. 4.13, $k_u = 0.01$, $f_{\text{ref}} = 10^7$ Hz, $n_{\text{fract}} \approx 0.015$, giving $\mathcal{L}_{\text{out},u3}(0.01) \approx -106$ dBc/Hz and $\mathcal{L}_{\text{out},u4}(0.01) \approx -102$ dBc/Hz.

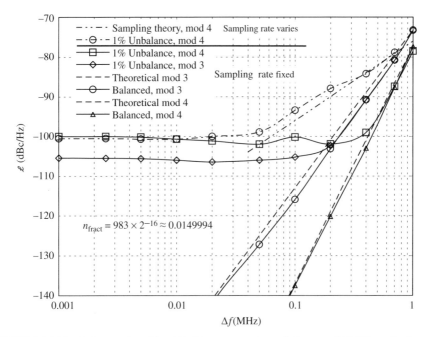

FIGURE 4.13 Noise produced by 1% CP current mismatch in some MASH synthesizers. $f_{\text{ref}} = 10\,\text{MHz}, f_{\text{out}} \approx 100.15\,\text{MHz}$. These curves do not show the attenuation due to the loop response that occurs as the higher values of Δf.

According to Eq. (4.8), the level $\mathcal{L}_{\text{flat}}$ represents a current mismatch, between the up and down current sources in the PD, of only 0.38% (at the worst value of n_{fract}), so it seems unlikely that the high-frequency noise floor of $\mathcal{L}_{\text{bj}}$ can be maintained in the presence of ΣΔ modulation.

We will consider this current mismatch problem further as we obtain the noise levels, due to the three effects considered here, for the three other loop configurations that we had considered in Section 4.1.6.

4.2.4 Levels for All Four Loop Configurations

Just as we used Fig. 4.4 to find the level of $\mathcal{L}_{\text{flat}}$ that will produce output PSD that, at its closest, is 6 dB below the maximum noise, so also we can determine this for the other three designs from Figs. 4.6, 4.8, and 4.9. For each design, the degree of current mismatch and the dither bit required to produce that level are each listed in Table 4.3 The margin by which the excess noise due to varying sampling rate is below the highest noise is also given for a chosen frequency. That frequency is chosen for the worst case (by considering how the loop affects that noise in comparison to the other curves in the figures). For example, in Fig. 4.12, that frequency is 141 kHz, where the varying sample rate noise protrudes the most above the VCO noise.

Table 4.3 and Figs. 4.1–4.12 (or variations of them) can be obtained by using `responses.xls`.

TABLE 4.3 Noise Levels Due to Defects

	A	B	C	D	E	F	G	H	I
						k_u ($n_{fract} \approx 1$)	maximum dither bit $q=1$	$\mathcal{L}_{out}$, with varying sample rate	
2	Refer to	*$\mathcal{L}_{out}$ level to compare							
3		10,100 MHz				0.38%	−9.044	−101.2 dB	before $|H(fm)|$
4	Fig. 4.1	* closed loop		−126 dBc				−121.729	after $|H(fm)|$
5	Fig. 4.4	@f_m =		1.41E+05 Hz				−4.3 dB	below other noise
6		$|H(fm)|$		−20.5 dB					(negative ⇒ above)
7		$\mathcal{L}_{flat}$ =		−111.5 dBc					
8		f_{ref}=		1.00E+07 Hz				out	
9		f_{out} =		1.00E+08 Hz					
10		10,1000 MHz				1.73%	−6.844	−114.2 dB	before $|H(fm)|$
11	Fig. 4.5	* closed loop		−113dBc				−135.320	after $|H(fm)|$
12	Fig. 4.6	@f_m =		3.16E+05 Hz				22.3 dB	below other noise
13		$|H(fm)|$		−21.1 dB					(negative ⇒ above)
14		$\mathcal{L}_{flat}$ =		−98.25 dBc					
15		f_{ref}=		1.00E+07 Hz					
16		f_{out} =		1.00E+09 Hz					
17		100,1000 MHz Narrow				5.46%	−5.186	−118.2 dB	before $|H(fm)|$
18	Fig. 4.7	* closed loop		−119 dBc				−139.253	after $|H(fm)|$
19	Fig. 4.8	@f_m =		6.31E+05 Hz				20.3 dB	below other noise
20		$|H(fm)|$		−21.0 dB					(negative ⇒ above)
21		$\mathcal{L}_{flat}$ =		−98.27 dBc					
22		f_{ref}=		1.00E+08 Hz					
23		f_{out} =		1.00E+09 Hz					
24		100,1000 MHz Wide				1.12%	−7.467	−108.2 dB	before $|H(fm)|$
25	Fig. 4.7	* closed loop		−126 dBc				−128.153	after $|H(fm)|$
26	Fig. 4.9	@f_m =		2.00E+06 Hz				2.2 dB	below other noise
27		$|H(fm)|$		−20.0 dB					(negative ⇒ above)
28		$\mathcal{L}_{flat}$ =		−112 dBc					
29		f_{ref}=		1.00E+08 Hz					
30		f_{out} =		1.00E+09 Hz					
31		order:		3	⇐Use only 3 or 4 for k_u ; also 5 for $\mathcal{L}_{out}$.				
32		*Choose f_m and a "closed loop" $\mathcal{L}_{out}$ for comparison with $\mathcal{L}_{out}$ due to varying sampling rate.							

We see that dithering of the tenth least significant bit would generate a noise level too low to be seen in any of the loops (if it were processed as we have discussed).

Excess noise due to varying sampling rate is below total noise in these additional three loops. It is farthest below in the 10, 1000 loop. The other loops also permit greater charge pump mismatch before an effect can be seen. The key to these variations is the parameter dependencies shown in Table 4.1.

4.2.5 Simple Charge Pump

Now let us consider the noise due to CP current mismatch a bit further. Our previous simulation employed ideal current sources so the variation in currents would be due to some matching problem, perhaps temperature related. However, severe current mismatch can occur due to finite CP output impedance if the CP drives a varying voltage.

The simplest charge pump consists of a pair of transistor switches that connects a resistor to the positive or negative supply, which we model as shown in Fig. 4.14. The capacitor in Fig. 4.14 is part of the loop filter and performs integration. An op-amp integrator (e.g., Fig. F.6.2e, which includes the resistor) would present a fixed voltage to the resistor, but in this simple circuit that voltage varies as the capacitor voltage changes to accommodate various tuned frequencies, and this does produce distortion, increasing or decreasing the current produced in response to U relative to that produced in response to D. (The lead circuit, which is required for stability, was part of the rest of the filter in this experiment; see Section H.2.)

Results from simulations are shown in Fig. 4.15. Curve 6 is the expected $\mathcal{L}(f_m)$ from Eq. (2.3) for this MASH-111 synthesizer, but the measured noise levels off at various floors. Curves 4 and 5 are floors that occur with two different capacitor values when the PD output voltage is in the center of its range at 2.5 V. In that case, the pulse current is 2.5 V/520 Ω = 4. 8 mA. The PD pulse width due to a single swallow at 100.15 MHz output frequency is 10^{-8} s and this produces a voltage change of

$$\frac{(4.8 \times 10^{-3}\,\mathrm{C/s}) \times 10^{-8}\mathrm{s}}{2.2 \times 10^{-7}\,\mathrm{C/V}} = 0.22\,\mathrm{mV}, \tag{4.10}$$

with the 0.22 μF capacitor (curve 5) or 2.2 mV with the 0.022 μF capacitor (curve 4). The voltage across the 0.22 μF capacitor is shown in Fig. 4.16. Note that the peak value of the waveform is also about 0.2 mV. This does not change much with f_{fract} (the variance of the sequence shows little dependence on n_{fract}). The resulting relative change in pulse current, and thus K_p, equals this voltage divided by 2.5 V (the voltage

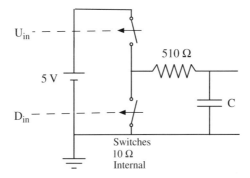

FIGURE 4.14   Simple CP PD. Transistors are represented as switches driven by digital inputs, U_{in} and D_{in}.

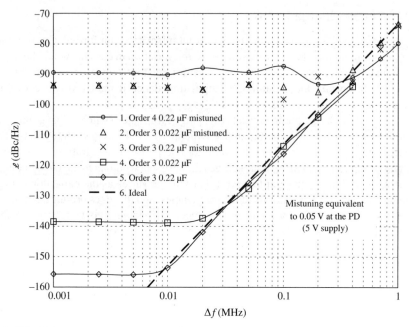

FIGURE 4.15 Noise floors with a simple charge pump where the current pulses are affected by the charge on the following integrating capacitor. These curves do not show the attenuation due to the loop response that occurs as the higher values of Δf.

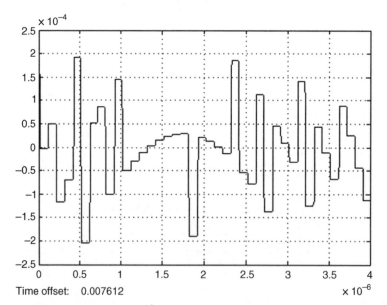

FIGURE 4.16 Waveform on $0.22\,\mu F$ capacitor. The CP current pulses occur at the beginnings of the pulses shown and are responsible for the voltage changes on the capacitor. $f_{out} = 100.15$ ... MHz, $f_{ref} = 10$ MHz.

across the resistor), about 10^{-4} and 10^{-3}, respectively. Yet this small modulation of the sequence, from sample to sample, creates the noise levels shown by curves 5 and 4 in Fig. 4.15.

As the output frequency changes, so does the average voltage on the capacitor. Curves 2 and 3 show the effects of this "mistuning," which amounts to 50 mV, considerably larger than the variations due to the sequence. This mistuning represents 86 kHz change in f_{out} with the smaller capacitor and 860 kHz with the larger one (the gain from the capacitor is changed to maintain the same overall loop transfer function with different capacitors). It causes a finite, relatively constant, ratio between positive and negative current amplitudes. The resulting errors in charge delivered to the capacitors, and the corresponding phase, is independent of the capacitor size, as is the resulting noise floor. The mistuning here is relatively small compared to the range required in many synthesizers, so it is obvious that the noise could be larger. Since the frequency change corresponding to a given voltage offset is proportional to the size of the capacitor, the importance of larger capacitors is apparent.

Curve 1, compared to curves 2 and 3, shows the increase in noise level with an increase in the order of the $\Sigma\Delta$ modulator. That ratio between noises for the two orders is similar to what was seen with ideal CPs (Fig. 4.13). It is due to the resulting increase of sequence amplitude and thus phase deviation. Noise produced by nonlinearities of various kinds, such as those to be described below, is commonly worse for higher orders because of the greater deviations of the sequence.

Since no current source has infinite output impedance, it is likely that these effects will be seen even with the sorts of current sources that are usually employed, if they drive a varying voltage. In contrast to our treatment of the more ideal charge pump, we do not provide a formula for computing the magnitude of the effect, although the Simulink® model is available and can be adapted.

4.2.6 System Performance

We have studied how to prevent the defects from increasing the level of $\mathcal{L}_{out}$ significantly as a way to gain insight into the noise sources and their variations with parameters. However, in most system designs, the objective will be to prevent these various noise levels from negatively impacting the system performance. Any of the noise levels that we have considered could be excessive for that purpose or they may be allowable, even though they increase total noise.

4.3 NOISE SOURCES: EQUIVALENT INPUT NOISE

Noise sources have been identified that have a theoretical dependence on f_{ref} and on f_m that is the same as that in the observed noise described as $\mathcal{L}_{bj}$ by Eq. (4.6), so they may explain the source of that noise. It is important to understand these sources not only for design and troubleshooting but also because some of them show a dependence on $\Sigma\Delta$ modulation, which is not included in Eq. (4.6), and therefore portend a degradation in the presence of such modulation.

4.3.1 Without ΣΔ Modulation

Since, in general, phase jitter is related to timing jitter by

$$\delta\varphi = f\delta t, \qquad (4.11)$$

a timing jitter density $S_{\Delta t}$ leads to a PPSD, all referred to the loop input, of

$$S_{\varphi,\text{ref}}\big|_{\text{jitter}} = f_{\text{ref}}^2 S_{\Delta t}, \qquad (4.12)$$

but if this noise is sampled (see Appendix N), as it occurs when it produces a logic change (we can only tell the phase at that moment), the noise will be replicated in the frequency domain at each multiple of the sampling frequency f_{ref}. Then, assuming that the sampling "instant" (the rise time of the logic) is short,[22] the number of noise overlaps at low frequencies (where we are most concerned) will be approximately[23] equal to the number of times that f_{ref} fits into the width of the two-sided noise band $w_{2\text{nb}}$. If the jitter noise spectrum is flat, this results in a sampled value of phase noise that is proportional to $1/f_{\text{ref}}$, changing Eq. (4.12) to

$$S_{\varphi,\text{ref}}\big|_{j,\text{samp}} = S_{\varphi,\text{ref}}\big|_{\text{jitter}} \frac{w_{2\text{nb}}}{f_{\text{ref}}} = w_{2\text{nb}} S_{\Delta t} f_{\text{ref}}. \qquad (4.13)$$

Brennan and Thompson [2001] found this frequency dependency to exist in PFDs on which they performed measurements and ascribed the noise to sampled jitter.

The white phase noise in Eq. (4.6) has the same dependency on f_{ref} as the sampled jitter in Eq. (4.13), while the flicker noise has the same dependency as the unsampled jitter in Eq. (4.12). This is not surprising because the broadband white noise is increased due to the addition of aliased copies, but the flicker noise is not so affected since flicker noise spectrums that are centered at multiples of f_{ref} are relatively weak near zero frequency. Thus, we may write Eq. (4.6), for large charge pump currents $(I_{\text{cp}} \gg I_{\text{knee}})$, as

$$\mathcal{L}_{\text{ref,bj}}\big|_{I_{\text{cp}} \gg I_{\text{knee}}} = w_{2\text{nb}} S_{\Delta t}\big|_{\text{white}} f_{\text{ref}} + \frac{S_{\Delta t}(f_m = 1\ \text{Hz})\big|_{\text{flicker}}}{f_m/\text{Hz}} f_{\text{ref}}^2. \qquad (4.14)$$

However, phase noise produced by these time jitter-related noises does not depend on I_{cp}, so when $I_{\text{cp}} \ll I_{\text{knee}}$ and therefore Eq. (4.6) becomes proportional to $1/I_{\text{cp}}$, they no longer explain Eq. (4.6). Under this condition, we transition to an expression, from Arora et al. [2005, last equation in Section IV], for noise originating in the CP current noise. We simplify their equation by assuming that the noise PSDs from the up and down pumps are the same, both $S_{i,\text{cp}}$, and obtain

$$S_i(f_m)\big|_{\text{cp,samp}} = 2S_{i,\text{cp}}\left\{1 + \frac{f_c}{f_m}\left(\frac{\sigma_{\delta k}/\sqrt{2\pi} + T_{\text{delay}}}{T_{\text{ref}}}\right)\right\}\left(\frac{\sigma_{\delta k}/\sqrt{2\pi} + T_{\text{delay}}}{T_{\text{ref}}}\right), \qquad (4.15)$$

where f_c is the flicker corner frequency, $\sigma_{\delta k}$ is the standard deviation of the time of arrival due to the $\Sigma\Delta$ sequence, and T_{delay} is the delay built into the PFD to ensure that both charge pumps go on at zero phase error (see Appendix P and Figs. F.5.C.10 and F.5.C.13). $S_i(f_m)|_{\text{cp,samp}}$ is the noise from $S_{i,\text{cp}}$ after it has been sampled by the pulse train. Its noise level in the flat region has been increased by overlapping aliased noise, but the flicker noise is not so affected.

If we assume that the charge pump current I_{cp} is generated from a resistor with a voltage V across it, the power spectral density will be

$$S_{i,\text{cp}} = 4N_T/R = 4N_T I_{\text{cp}}/V. \tag{4.16}$$

We can now convert the current PSD in Eq. (4.15) to phase PSD by multiplying it by the square of

$$1/K_p = 2\pi \text{ rad}/I_{\text{cp}}. \tag{4.17}$$

Combining the last three equations, we obtain

$$S_{\varphi,\text{ref}}|_{\text{cp,samp}}(f_m) = \frac{32\pi^2 N_T}{I_{\text{cp}} V} \left\{ \begin{array}{l} f_{\text{ref}}(\sigma_{\delta k}/\sqrt{2\pi} + T_{\text{delay}}) \\ + f_{\text{ref}}^2 \dfrac{f_c}{f_m}(\sigma_{\delta k}/\sqrt{2\pi} + T_{\text{delay}})^2 \end{array} \right\}. \tag{4.18}$$

We can see from this expression how the noise is increased by the delay that is used to prevent nonlinearity near zero phase. The expression has the same dependence on I_{cp} (as well as on f_{ref} and f_m) as does Eq. (4.6). Thus, we could explain Eq. (4.6) as a sum of PSDs given by Eq. (4.14) and by Eq. (4.18) with $\sigma_{\delta k} = 0$ (i.e., no $\Sigma\Delta$ modulation). The latter dominates at low CP currents because while the noise power is proportional to the current, K_p is also proportional to it, and the equivalent PSD is proportional to $1/K_p^2$.

4.3.2 Increase with $\Sigma\Delta$ Modulation

Significantly, we also see that $S_{\varphi,\text{ref}}|_{\text{cp,samp}}(f_m)$ depends on the variance of the integrated $\Sigma\Delta$ sequence since $\sigma_{\delta k} = \sigma_{\Sigma\delta N}/f_{\text{out}}$, so our equivalent input noise is increased in the presence of $\Sigma\Delta$ modulation, and more so for higher orders. Since the digital integration of the sequence basically reduces its order by 1 [Eq. (M.26)], we can write

$$\sigma_{\delta k}(p) = \frac{\sigma_{\delta N}(p-1)}{f_{\text{out}}}. \tag{4.19}$$

Section M.7 gives $\sigma_{\delta N}(p = 2) = 0.5$ and $\sigma_{\delta N}(p = 3) = 1.67$, so in Eq. (4.18), $\sigma_{\delta k}/\sqrt{2\pi} = 0.2/f_{\text{out}}$ for MASH-111 and $\sigma_{\delta k}/\sqrt{2\pi} = 0.67/f_{\text{out}}$ for MASH-1111. These values are to be compared with T_{delay} in each case. Since both T_{delay} and

$1/f_{out}$ need to be a little larger than rise times, they are likely to be of the same order of magnitude, so ΣΔ modulation may make only a small increase in the noise, but that is design dependent.

Arora et al. [2005] have also found that various timing imperfections can produce noise floors in the presence of ΣΔ modulation, so the equivalent input noise can increase with modulation even at high current levels. Such noise floors can be produced by finite rise and fall times of the charge pump, or if, in the PFD (Fig. F.5.20), the reset delays (from R) of the bistables are influenced by the polarity of their clocks (RD and VD), or if delays in the frequency divider are influenced by its modulus. All the resulting PSDs, reference to the loop input, are proportional to f_{ref},[24] the same dependency possessed by the noise floor in Eq. (4.6), so k_1 is subject to increase with ΣΔ modulation. The noise due to finite rise and fall times is also greater for higher order modulators.

Since the noise floor at both high and low currents is subject to increase with ΣΔ modulation, we would expect I_{knee} to also change with modulation. Therefore, it would be helpful to have the parameters of Eq. (4.6) not only for different ICs but also with ΣΔ modulation, and even for different modulators (e.g., for MASH-111 and MASH-1111). While we have not identified any modulation-dependent flicker noise, it seems likely that k_1 and k_{11} would begin to be affected by low I_{cp} values at different points, at least under some conditions. If the flicker noise does not change, a change in I_{knee} would require a change in k_{11} in order to maintain it constant anyway. Therefore, a more general form of Eq. (4.6) may be written as

$$\mathscr{L}_{ref,equiv}(M) = k_1(M)f_{ref}\left[1 + \frac{I_{knee,1}(M)}{I_{cp}}\right] + k_{11}(M)\frac{f_{ref}^2}{f_m}\left[1 + \frac{I_{knee,11}(M)}{I_{cp}}\right] \quad (4.20)$$

to indicate that k_1 and k_{11} may have different values of I_{knee} and that the various parameters can change with different modulators M.

If it turns out that Eq. (4.6) is an accurate representation for a given IC, independent of ΣΔ modulation, that too is important information.

4.4 DISCRETE SIDEBANDS

4.4.1 At Offsets Related to f_{fract}

4.4.1.1 *Due to Current Mismatch*   Several researchers[2] have observed that mismatched PD currents create a nonlinearity that causes discrete spurs at offsets equal to f_{fract} and harmonics thereof. We can observe spurious sidebands related to the mismatch, at frequency offsets equal to the reciprocal of the sequence duration, in simulations with a CP PD, simulations such as those used to produce Fig. 4.13 (but with n_{fract} set to 0.0625), even with a MASH-111 modulator initialized to 2^{-16} (e.g., with 5% mismatch, -59 dBc at 0.5 f_{fract} and -79 dBc at f_{fract} and 1.5 f_{fract}). The nonlinearity arises from the fact that the charged delivered by a positive pulse of width ΔT is

$$Q_+ = \left(I + \frac{\Delta I}{2}\right)\Delta T = \left(I + \frac{\Delta I}{2}\right)\frac{\Delta\varphi}{f_{\text{ref}}} \qquad (4.21)$$

when $\Delta\varphi$, the phase difference between reference and divider outputs, is positive, whereas the negative pulse delivers

$$Q_- = \left(I - \frac{\Delta I}{2}\right)\Delta T = \left(I - \frac{\Delta I}{2}\right)\frac{\Delta\varphi}{f_{\text{ref}}} \qquad (4.22)$$

when $\Delta\varphi$ is negative. Therefore, the charge delivered is

$$Q = I\Delta T + \frac{\Delta I}{2}|\Delta T| \qquad (4.23)$$

$$= \frac{\Delta\varphi}{f_{\text{ref}}}\left(I + \frac{\Delta I}{2}\,\text{sign}\,(\Delta\varphi)\right). \qquad (4.24)$$

This nonlinearity is not caused by a varying sampling rate, so resampling does not eliminate it.

Pamarti et al. [2004] developed a circuit to cancel this nonlinearity. In their circuit, each charge pump current is shared between two identical current sources, one of which turns off at a fixed delay after the second PFD input, as is typical, while the other at a fixed delay after the first input. The cancellation occurs because the current from the added charge pumps equals $\Delta I/2$ during the time both positive and negative pumps are on, but that duration shortens as $|\Delta T|$ increases, just countering the nonlinear term in Eq. (4.23). The two half-current pumps of *each* polarity must be equal and remain equal, even though we are assuming a difference between the pumps of *different* polarities, perhaps due to a greater difficulty in matching the latter. The solution also applies where the cause of a current variation affects both pumps of the same polarity the same, as does variation of the output voltage with finite source impedance in Section 4.2.5.

4.4.1.2 *Not Necessarily Related to Mismatch*   We encountered spurs at offsets equal to f_{fract} and its harmonics and subharmonics in Chapter 2. Banerjee [2008] has observed that some spurs at offsets of f_{fract} and its harmonics, in ICs, act as though they were not passing through the loop filter but rather influencing the VCO by some other path. He has, therefore, called these "crosstalk spurs" and combined their predicted power, at any spur frequency, with that predicted due to signals passing through the loop filter. Of course, crosstalk in general is a problem in frequency synthesizers (Section F.3.9.8) and the VCO is susceptible to coupling via its output port (Sections F.3.6.2 and F.8.1.1) and to coupling via grounds and power supply lines.

4.4.2 At Offsets of nF_{ref}

The Fourier series for waveforms that repeat with a period $1/F_{\text{ref}}$ contains harmonics of F_{ref}. When they get on the tuning voltage, they cause sidebands on the output spectrum that are separated from the main spectral line at F_{out} by multiples of F_{ref}. We call these reference–frequency spurs or reference spurs. Such repetitive waveforms arise from frequency divider and phase detector waveforms, especially from the latter since they are in the control–voltage signal path. They are discussed extensively in Chapter F.5. Here, we will describe reference sidebands that are caused by ΣΔ modulation and how they interact with two other types of reference sidebands. We will concentrate on the first sideband, which is usually of most interest because it receives the least attenuation from low-pass filtering. Other nearby sidebands at offsets of nF_{ref} can be treated similarly as long as the pulses that cause them are narrow compared to $1/(nF_{ref})$. Derivations and supporting data from simulations are provided in Appendix R. Initially we assume a varying sample rate (no resampling).

4.4.2.1   *Due to ΣΔ Modulation*   The ΣΔ modulator produces a series of numbers $n_c(i)$ that determine how many swallows occur in the frequency counter. This in turn determines the phase modulation produced at the PD. At some reference period i, the divider output is time modulated by

$$\Delta t(i) = n_d(i)/F_{out}, \tag{4.25}$$

where n_d is the digital integral of the carry sequence,

$$n_d(z) = \frac{n_c(z)}{1-z^{-1}}. \tag{4.26}$$

Because the positive and negative pulses occur on different sides of the reference transition (see "pump up" and "pump down" logic signals in Fig. 2.6), sinusoidal components at harmonics of F_{ref} are created. The component at the fundamental frequency F_{ref} is equivalent to a sinusoid of phase at the loop input (the reference input) with peak phase deviation [Eq. (R.13)]

$$m_{1,ref,\Sigma\Delta} = \left(\frac{2\pi}{N}\right)^2 \sigma_{nd}^2, \tag{4.27}$$

where σ_{nd}^2 is the variance of the sequence n_d. Based on simulations, the deviation for a third-order MASH is [Eq. (R.15)]

$$m_{1,ref,\Sigma\Delta3} = 19.7 \, rad/N^2, \tag{4.28}$$

while that for fourth order is [Eq. (R.16)]

$$m_{1,ref,\Sigma\Delta4} = 65.8 \, rad/N^2. \tag{4.29}$$

The corresponding sideband level at the output would be obtained as[25]

$$SB_{1,out,\Sigma\Delta} = m_{1,ref,\Sigma\Delta}NH(F_{ref})/2. \tag{4.30}$$

These equations also apply to other modulators that produce the same output as the MASH modulators.

4.4.2.2 Due to Delays in the PFD   It is common for a PFD to generate two equal and opposite pulses at zero phase in order to avoid crossover distortion (Section F.5.7.2). Unequal circuit time delays can cause one of these pulses to end before the other and the PLL would then cause the other pulse to start sooner, leading to two equal-width pulses that are offset in time by T_o. Such offsets lead to equivalent input phase modulation given by [Eq. (R.8)]

$$m_{1,\text{ref},o} = 8\pi^2 \text{ rad} D_p D_o, \tag{4.31}$$

where

$$D_o = T_o/T_{\text{ref}}, \tag{4.32}$$

D_p is the duty factor of each pulse, and T_o is the offset between them. This would again produce an output sideband level obtained as in Eq. (4.30).

Current imbalance between opposing pulses also leads to offset (Section F.5.7.6). Sidebands produced by the resulting offset are more significant than those produced by the current imbalance.

Since the sinusoids producing the sideband both due to the delay and due to $\Sigma\Delta$ modulation pass through zero approximately at the reference transition, they will be in phase or opposed. Therefore, when both effects are present, we have

$$m_{1,\Sigma\Delta o} = m_{1,\Sigma\Delta} \pm m_{1,o}, \tag{4.33}$$

with the sign depending on which delay in the logic is the longer.

4.4.2.3 Due to Leakage Current   Leakage current I_L at the charge pump output leads to [Eq. (R.5)]

$$m_{1,\text{ref},L} = 2I_L/K_p, \tag{4.34}$$

where the output level can be obtained again as in Eq. (4.30). Unlike the previous two cases, this sinusoid peaks near the reference transition and as such tends to be in quadrature to the other two.

4.4.2.4 Due to All Three   If all three effects were present, the level would be given by

$$m_{1,\Sigma\Delta Lo}^2 = (m_{1,\Sigma\Delta} \pm m_{1,o})^2 + m_{1,L}^2. \tag{4.35}$$

We note that the three values of $m_{1,\text{ref}}$ depend differently on the loop parameters: $m_{1,\text{ref},\Sigma\Delta} \sim 1/N^2 = F_{\text{ref}}^2/F_{\text{out}}^2$, $m_{1,\text{ref},o} \sim F_{\text{ref}}^2$,[26] and $m_{1,\text{ref},L}$ depends on neither. All the modulation indexes that are referred to the input (i.e., m_{ref}) must be multiplied by $NH(F_{ref})/2$ to obtain the sideband level, assuming $m \ll 1$, as in Eq. (4.30).

4.4.2.5 *With Resampling*   If the PD output is resampled (e.g., Figs. 2.5 or 2.23), none of the waveforms described above will be seen on the tuning line and none of these sidebands will be produced. However, the sampler will inevitably produce some level of transient at an F_{ref} rate due to capacitive coupling of the controlling signal, for example; so sidebands from this effect are possible.

4.4.2.6 *Significance of Levels*   Are the levels of the reference sidebands significant and, if so, under what conditions?

The output reference spur levels that occur in our models, even when the filtering has been reduced to help us see the spur on the tuning voltage, are very small, but we may want to use even less filtering in some applications. In a fractional-N synthesizer, the filtering requirement may be driven by the spur levels or it may be driven by the quantization noise, so it is worth considering how the level of these two undesired sidebands compare. Since there is likely to be some similarity in the effects of the two noise powers, we can usefully compare them. We will compare levels referenced to the output before attenuation by $H(f_m)$.

The quantization noise peaks at $F_{ref}/2$ and in that region ($f_m < F_{ref}$), it has a noise bandwidth,[27] for third- and fourth-order modulators, of 0.38 F_{ref} and 0.31 F_{ref}, respectively. Thus, most of the quantization noise will be near $F_{ref}/2$, so there will be more attenuation for the reference sideband at $f_m = F_{ref}$ than for most of the quantization noise below F_{ref}. The difference depends on the order of the filter, but since the frequencies of importance, F_{ref} and $F_{ref}/2$, are only one octave apart, it will be limited. For third- and fourth-order modulators, respectively, the integrated quantization noises referred to the output, before attenuation by the loop response, have powers, relative to the main signal, of [28]

$$\int_0^{F_{ref}} \frac{\mathcal{L}_\varphi(p=3)}{|H(f_m)|} \, df_m = 12.95 \text{ dBc} \qquad (4.36)$$

and

$$\int_0^{F_{ref}} \frac{\mathcal{L}_\varphi(p=4)}{|H(f_m)|} \, df_m = 18.18 \text{ dBc}. \qquad (4.37)$$

Here, $\mathcal{L}_\varphi = S_\varphi/2 \neq L_\varphi$ because the small modulation index requirement has definitely not been met. However, by the time the sidebands are reduced by filtering, in a practical case, $\mathcal{L}_\varphi$ will equal L_φ. We are using f_m and Δf interchangeably.

Curves 1 and 2 in Fig. 4.17 show the reference spur levels due to ΣΔ modulation [Eq. (4.30)] divided by $H(F_{ref})$] relative to these integrated quantization noise sidebands (without defects). We see that the discrete sideband power comes closest

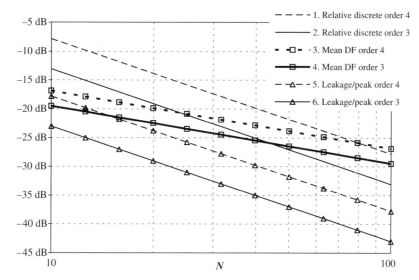

FIGURE 4.17 Comparison of discrete sidebands with integrated quantization noise sidebands. Curves 1 and 2 give the power of the discrete sidebands relative to the corresponding integrated noise. Curves 3–6 show the magnitude of certain defects that would cause the sideband power due to them to equal the integrated noise. Powers are reference to the output before reduction by $H(f_m)$.

to the integrated power for low values of N and for order 3. Even for $N = 10$, however, it is about 8 dB weaker.

Curves 3 and 4 give the value of $20 \log_{10} \sqrt{D_p D_o} = 10 \log_{10} D_p D_o$ [Eq. (4.31)] that would cause the sideband power due to offset pulses to equal that of the integrated quantization noise. Here we are most concerned about low values, which would indicate that the geometric mean of the two D values would need to be small in order to make the spurs insignificant. The worst case shown in Fig. 4.17 is -30 dB at $N = 100$ for the third-order modulator, implying that $D_p D_o$ should be small compared to 0.001. If the cancelling pulses in the PD are 1% of T_{ref} in width and their offset is a tenth of their width, $20 \log_{10} \sqrt{D_p D_o} = -50$ dB, so this requirement does not seem very difficult. Perhaps at high reference frequencies, it may become more difficult to keep the pulse that narrow relatively.

Curve 6 for order 3 at $N = 100$ indicates

$$20 \log_{10} I_p / I_L \ll -43 \mathrm{dB}$$

or $I_p / I_L \ll 7 \times 10^{-3}$, which means that the leakage must be smaller than 7 μA if the pulse amplitude is 1 mA. Again, this does not seem very severe. These curves are descending at -20 dB/decade, so the requirements become more difficult at higher N.

It seems that the requirement to filter the quantization noise makes the presence of these discrete sidebands less important than it would be in integer-N synthesizers.

4.4.3 Charge Pump Dead Zone

It is very important to avoid crossover distortion (dead zone) with ΣΔ synthesizers. Simulations have shown severe noise due to this effect [Crawford, 2008, pp. 374–377; De Muer and Steyaert, 2003, pp. 192–195]. With a charge pump PFD, a flat zone at zero phase of width k_d cycles can cause more noise than a current mismatch of $k_u = 100k_d.$[29] Crossover distortion can also cause large discrete spurs. These are in addition to the negative effects seen with integer-N synthesizers (Section F.5.7.2).

4.5 SUMMARY

- Both reference noise and component noise are referred to the input.
- Equivalent input noise density has been quantified in ICs.
  - It has flicker and white (flat) components.
  - It increases at low CP currents.
  - It can increase in the presence of ΣΔ modulation.
- Other noise densities occur with ΣΔ modulation:
  - theoretical quantization noise,
  - excess quantization noise from varying sample rate,
  - noise due to CP current mismatch,
  - LSB dither noise.
- These various noise densities are affected differently by loop parameters.
  - Total output noise depends on the noise sources and the loop.
  - They should be traded off to meet system requirements.
- Discrete sidebands can also be produced by ΣΔ modulation.
  - CP mismatch can lead to sidebands offset by multiples of f_{fract} or multiples of submultiples.
  - Variable sampling rate leads to sidebands offset by multiples of f_{ref}.
    - These combine with such sidebands due to other effects.
    - Filtering required for quantization noise may make these of less concern than in integer-N synthesizers.
  - Resampling eliminates all of these, but can also produce sidebands.
- Avoidance of PFD dead zone is very important.

CHAPTER 5

OTHER $\Sigma\Delta$ ARCHITECTURES

There are a large number of possible architectures other than MASH-11.... We will consider a few of these here.

5.1 STABILITY

One issue that we have not had to be concerned with in MASH configurations is the stability of the modulator. This is an issue for the more general $\Sigma\Delta$ modulator, where well-behaved performance is limited to a finite range of input values.

It should be apparent, with a little consideration, that the MASH modulator also has a limited range of input values. Fortunately, the allowed range is the range we require. Consider, however, what would happen if we input a number larger than 1. The first accumulator cannot output an average value larger than its maximum output, which is 1, and the outputs from the other accumulators are passed through difference functions $(1 - z^{-1})$, so their average contribution is 0.

The limits are not so obvious in more complex structures, where a single quantizer must output more than 1 bit. When their output ranges are restricted, they will perform properly only over a finite input range, which may not encompass the required range. We could observe the unrestricted output range over the full input range and provide for this in the design, but there may be reasons to restrict the output to a somewhat smaller range. For instance, increasing the range of the quantizer requires that the

Advanced Frequency Synthesis by Phase Lock, First Edition. William F. Egan.
© 2011 John Wiley & Sons, Inc. Published 2011 by John Wiley & Sons, Inc.

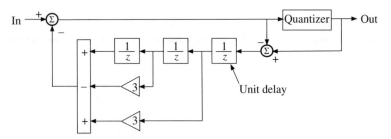

FIGURE 5.1   Model of third-order ΣΔ error feedback modulator.

pulse swallower also be designed for a larger range. If the quantizer range is smaller than the range that occurs without restriction, the amplitude at the input to the quantizer may build up, resulting in an increase in output noise or possibly a more severe failure. Crawford [2008, pp. 342–344] provides some design guidelines and recommends determining, by simulation, the relative occurrence of excessive values at the quantizer input, as the value of the modulator input is varied over its required range, in order to identify problem regions.

5.2 FEEDBACK[30]

One realization of the third-order error feedback modulator (EFM) configuration is shown in Fig. 5.1 (after Crawford [2008] who refers to Kozak and Kale [2003]) with the mathematical diagram shown in Fig. 5.2. From the latter, the output can be obtained as

$$n_{swallow} = n_{fract} + q(1-z^{-1})^3, \tag{5.1}$$

which is the same as MASH-111. To see this, consider that the transfer function from n_{fract} to $n_{swallow}$ is unity because the feedbacks from a and b cancel, producing zero net

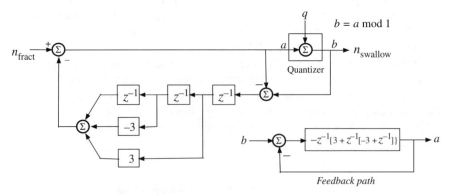

FIGURE 5.2   Mathematical model for Fig. 5.1.

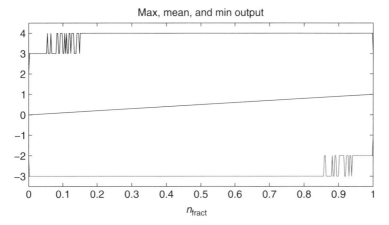

FIGURE 5.3   Maximum, mean, and minimum outputs from the error feedback modulator as a function of n_{fract}. The LSB is initially set in the first register in the feedback path.

feedback and leaving only the unity forward gain, and the transfer function from q to n_{swallow} is

$$\frac{b}{q} = \frac{1}{1 + G_R},$$

where G_R is given by the transfer function of the feedback path from b to a, shown in the lower right of Fig. 5.2. (The meaning of q is explained in Appendix M.)

Figure 5.3 shows the maximum, average (mean), and minimum output values from the modulator as a function of $x = n_{\text{fract}}$. The output ranges from -3 to $+4$, the same as the range of a MASH-111 modulator. Figure 5.4 shows the variance at the input and output of the modulator. The theoretical value at the output, 1.667, is given in Section M.7.

If we restrict the range to -2 to $+3$, the average output fails to equal n_{fract} over much of the range, as can be seen from Fig. 5.5. In these regions, the input to the quantizer has taken on very large values (Fig. 5.6) and the modulator has failed. (No real accumulator circuit is likely to accommodate such large values, so the failure would differ in detail.) The average output does appear to equal n_{fract} over the whole range when only one extreme is reduced, that is, for ranges of either -2 to $+4$ or -3 to $+3$. In these cases, there is up to 2% overdrive (i.e., clock cycles where the restriction affects the output at a given n_{fract}) and the output variance is reduced as much as 8% in the same regions, so the PPSD may also be affected there.

These plots[31] were produced by the MATLAB® script `FBall.m` (see Section 6.10). While the same data can be obtained from Simulink® models, the scripts are relatively easy to write for a purely digital modulator and they execute much faster.

There seems to be no significant difference between the performance of this configuration and MASH-111. The range of values of the carry and of the induced phase, as well as the statistics of these values, are about the same. The spectrum for the

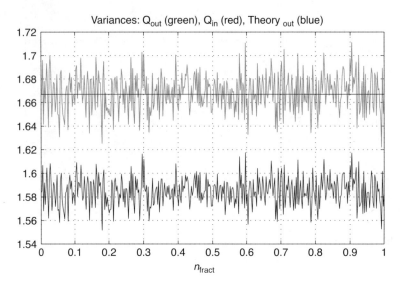

FIGURE 5.4   Variances of the input (lower trace) and output (upper trace) of the quantizer for error feedback modulator. The lower trace is σ^2 of Q_{in}.

EFM appears similar to that for MASH under the same conditions. Even the range of sequence lengths given by Borkowski et al. [2005] (see Section X.2) applies to the EFM as well as to MASH; a MASH modulator and an EFM with the same register lengths and order have the same range of sequence lengths. (They set only the first accumulator to an odd number in MASH, whereas other combinations are effective with the EFM, although this may offer no practical advantage.) In addition, the sequence length can be maximized by the same modulus change that is described in Section 3.2 [Hosseini and Kennedy, 2008]. The application of LSB jitter is discussed in Pamarti and Galton [2007] and the same restriction [Eq. (3.2)] applies that applies to MASH.

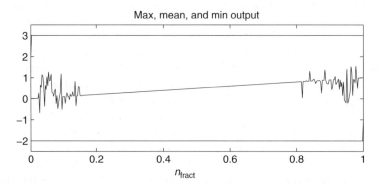

FIGURE 5.5   Outputs as in Fig. 5.3, but with output range limited to −2 to +3.

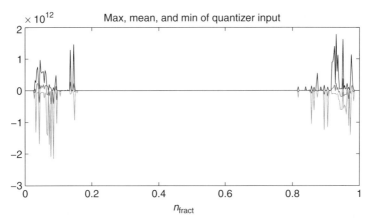

FIGURE 5.6   Inputs to the quantizer with output range limited to -2 to $+3$. (Note the 10^{12} multiplier for the y-axis.)

Other feedback structures can produce different transfer functions, as can the structure to be discussed next.

5.3 FEEDFORWARD[32]

Figure 5.7 shows an architecture that uses three feedforward paths and three accumulators but only a single quantizer and a single feedback path [Rhee et al., 2000]. We will call it the RS&A modulator. From the mathematical block diagram in Fig. 5.8, we can describe the output as

$$n_{\text{swallow}} = n_{\text{fract}} \frac{2z^{-1} - 2.5z^{-2} + z^{-3}}{1 - z^{-1} + 0.5z^{-2}} + q \frac{\left(1 - z^{-1}\right)^3}{1 - z^{-1} + 0.5z^{-2}}. \qquad (5.2)$$

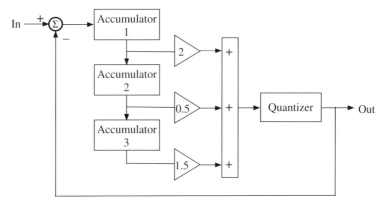

FIGURE 5.7   Model of third-order single-quantizer $\Sigma\Delta$ modulator using feedforward [Rhee et al., 2000].

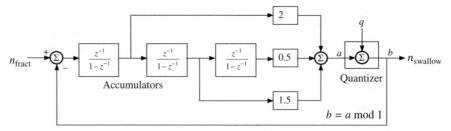

FIGURE 5.8 Mathematical model for Fig. 5.7.

The response to a step in n_{fract} is not an identical (if delayed) step, as it is for MASH, but it does settle to n_{fract}. The error, in response to a step, has a decaying exponential envelope with a time constant of about three sample periods, probably faster than that caused by the synthesizer's loop response.[33]

The numerator of the noise shaping function is the same as the entire function with MASH-111, but there are poles in the denominator that perturb the response to reduce the peak value at the expense of some more noise at lower frequencies, as shown in Fig. 5.9. The poles are an approximation of poles from a three-pole Butterworth filter, translated from the s-domain to the z-domain using the bilinear transformation.

The shape of this quantization noise may be advantageous in some applications and is one reason for choosing a particular form of modulator. In addition, the range of n_{swallow} from this circuit is considerably smaller than that from the MASH-111. This tends to reduce the negative effects of nonlinearities (Section 4.2.5) and possibly simplify the design of the multimodulus prescaler.

Figure 5.10 shows the output from the RS&A feedforward modulator. We can see that the output range, −2 to 3, is smaller than that of the MASH-111 or error feedback

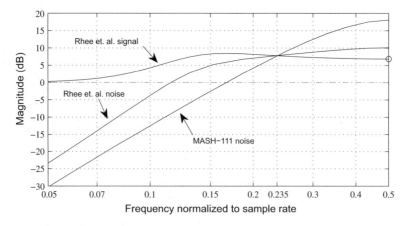

FIGURE 5.9 Responses at normalized modulation frequency of modulator from Rhee et. al. [2000] compared to MASH-111.

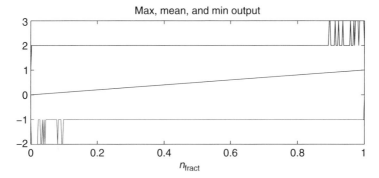

FIGURE 5.10 Unrestricted output from the feedforward modulator of Fig. 5.8.

modulators (-3 to 4 per Fig. 5.3). Moreover, when we restrict the output to an even smaller range of -1 to 2 (Fig. 5.11), it still outputs a mean value equal to n_{fract}. Figure 5.12 shows that the highest observed overdrive is less than 0.03% with the more severe limiting.

When we look at the variance of the modulator output (Fig. 5.13), however, we find something surprising. While its mean value is approximately equal to theory, as determined by integrating the quantization noise in Eq. (5.2), the variance changes slightly ($\sim$1 dB) with n_{fract}. (The plot looks the same for the unrestricted range.) A 1 dB difference in noise power does not stand out in a PPSD plot, but we can see some differences between superimposed plots taken with $n_{fract} = 2^{-15}$ and $n_{fract} = 0.5 + 2^{-15}$ in Fig. 5.14. The former is slightly higher at high frequencies and slightly lower at low frequencies, but the integral would be higher for the former, consistent with Fig. 5.13.

Simulink simulations[34] for the unrestricted output gave similar results to the MATLAB scripts and also showed a range at the phase detector (obtained by accumulating the modulator output) of 2.5 for the feedforward circuit compared to 3.9 for MASH-111, while the variances were approximately 0.2 and 0.5, respectively.

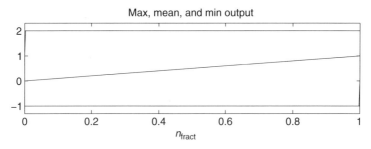

FIGURE 5.11 Output from the feedforward modulator of Fig. 5.8, restricted output range (-1 to 2).

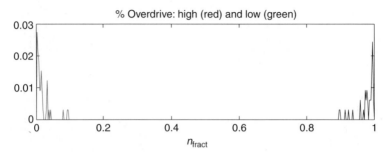

FIGURE 5.12   Overdrive for output of Fig. 5.11, low on left and high on right. This is the relative clock intervals during which the output is altered due to the range restriction.

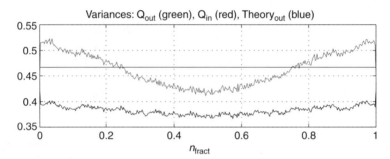

FIGURE 5.13   Variance of output (upper curve) and of the input to the quantizer (lower curve) corresponding to Figs. 5.10–5.12.

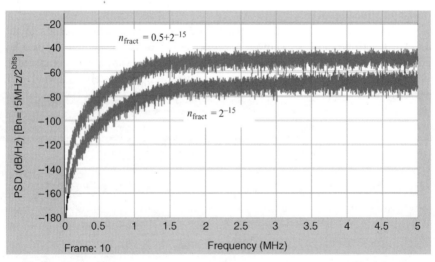

FIGURE 5.14   Modulator output PSDs taken near $n_{\text{fract}} = 0.5$, shifted up 20 dB, and near $n_{\text{fract}} = 0$. The lower curve seems to have higher PSD on the right, while the upper curve is higher on the left.

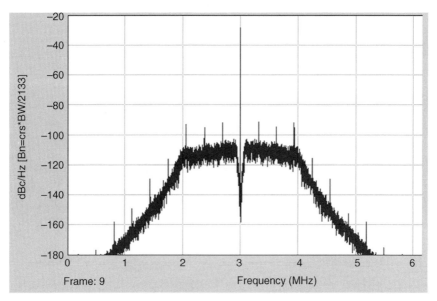

FIGURE 5.15 Output spectrum at 100.625 MHz with third-order single-quantizer loop [Rhee et. al., 2000] and 2^{-16} seed in first accumulator. $B_n = 703$ Hz.

However, the spectrum with $n_{\text{fract}} = 0.0625$ (Fig. 5.15) shows noticeable spurs at offsets that are multiples of 312.5 kHz compared to the spectrum for MASH-111 (Fig. 5.16). When n_{fract} was increased by 2^{-16}, with or without an initial 2^{-16}, many of the spurs disappeared. However, spurs offset by 2 and 3 times f_{fract} persisted. Multiple spurs were also present in simulations with $n_{\text{fract}} = 0.25 + 2^{-7}$ but they largely disappeared when n_{fract} was increased by 2^{-16} (i.e., $n_{\text{fract}} = 0.0100001000000001_2$), the setting employed by Rhee et al.[35] Figure 5.17 shows the two spectrums on the same log plot in the region where, according to Fig. 5.9, noise from the MASH configuration is lower.

Feedback and feedforward can both be used in the same modulator [Crawford, 2008, pp. 359–363; Rogers et al., 2006, pp. 330–337].

5.4 QUANTIZER OFFSET

We have not discussed details of the quantizer in our models. The question arises, should the quantizer be midtread or midriser (e.g., Fig. 6.12). That is, for example, should the output be 0 over an input range of -0.5 to $+0.5$ or over 0–1 (as in the first-order MASH modulators). It turns out that it does not make much difference. The loops adjust the input range to the quantizer to compensate. In fact, significant offsets can occur in series with the quantizer input without affecting performance. (The change from midrise to midtread is equivalent to an offset of 0.5. See Section 6.10 for further discussion.) The range of outputs from the registers will adjust to compensate. Of course, this means that the range required from the registers depends on the offset.

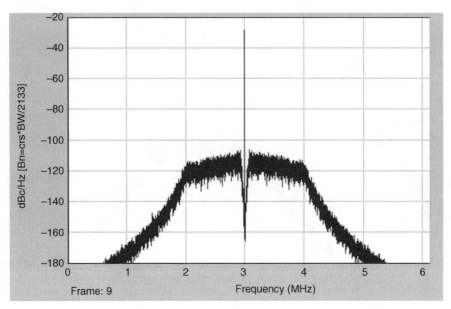

FIGURE 5.16 Output spectrum at 100.625 MHz with MASH-111 and 2^{-16} seed in first accumulator. $B_n = 703$ Hz.

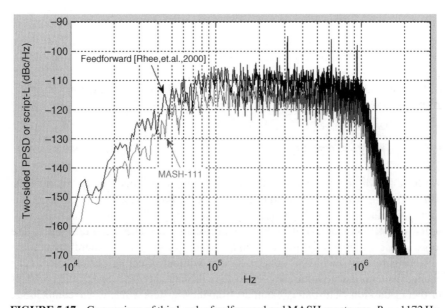

FIGURE 5.17 Comparison of third-order feedforward and MASH spectrums. $B_n = 1172$ Hz.

In addition, if the initial settings in the registers differ too much from the final values to which they would adjust, the adjustment, and the modulator, can fail (at least this is seen with the models where the registers have unlimited capacity).

Figure 5.18 is a plot showing the minimum, mean, and maximum inputs to the quantizers in the EFM and in the RS&A feedforward modulators as the offset is adjusted with a modulator input of $n_{\text{fract}} = 0.0625$. Zero offset corresponds to a midtread quantizer. Note how the maximum, mean, and minimum *inputs* to the quantizer adjust to compensate for the offset. In all cases, the mean *output* equaled the modulator input 0.0625. The minimum and maximum modulator outputs were also the same for each offset, -3 and $+3$ for the feedback modulator and -1 and $+2$ for the feedforward modulator, and the output variance was independent of offset for both modulators.

The offset range shown was the extent of the variation over which the modulators worked. Offsets outside that range (by 0.5) caused the quantizer inputs to become very large, while the outputs eventually pinned at a limit.

5.5 MASH-$n_1n_2n_3$[36]

Higher order $\Sigma\Delta$ modulators can be included in a MASH structure if the individual modulators have a forward transfer function consisting of just a delay z^{-m} and a transfer function from the quantization noise of $(1 - z^{-1})^n$, where n is the order of the

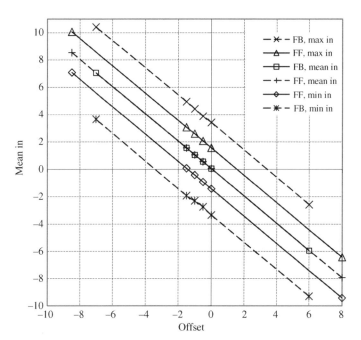

FIGURE 5.18   Input to the quantizer with 0.0625 into the modulator for EFM (FB) and RS&A (FF) modulators and 2^{-16} seed in the first register (others initially 0).

individual modulator. Then the configuration in Fig. 5.19 will cause all the quantization noises, except the last (q_3), to be canceled and the synthesizer's divider ratio
will be

$$N = N_{\text{int}} + n_{\text{fract}} z^{-m_1 - m_2 - m_3} + q_3 (1 - z^{-1})^p, \tag{5.3}$$

where the order of the MASH-$n_1 n_2 n_3$ is $p = n_1 + n_2 + n_3$.

5.6 CANCELLATION OF QUANTIZATION NOISE IN THE GENERAL MODULATOR

We have shown how to obtain a signal to cancel quantization noise in simple MASH
structures, but the general process, applicable to all modulators, is to
- subtract the modulator output from the n_{fract} input in order to obtain the negative
  of the quantization noise,
- divide by $1 - z^{-1}$, in order to duplicate the processing that occurs between the
  change of N and the phase error,
- effectively multiply by $K_p / \bar{N}$ to change the reference from the synthesizer output
  to the PD output, and
- match clock delays as required.

The cancellation that we have considered with MASH-11... modulators took advantage of some of these processes that occur within the modulator. Figures 1 and 4 in

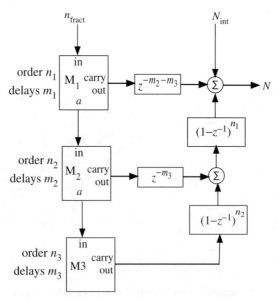

FIGURE 5.19 Connections for MASH-$n_1 n_2 n_3$.

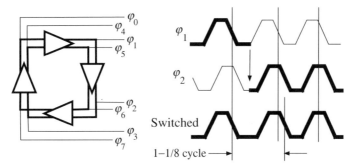

FIGURE 5.20 Phase switching with a four-element ring oscillator.

Pamarti et al. [2004] show how this process was accomplished with a single-quantizer second-order $\Sigma\Delta$ modulator while incorporating many of the techniques discussed in Section 2.1.3.

5.7 FRACTIONAL SWALLOWS

In the $\Sigma\Delta$ synthesizers that we have considered, the "swallow" or step in phase amounts to a whole number of cycles at f_{out}. A number of architectures have been proposed in which only a fraction of a cycle is swallowed. This can be accomplished by shifting the phase of the signal that drives the divider by a fraction of a cycle, thus reducing the quantization and the quantization noise by the same fraction. The use of ring oscillators (Section F.3.6.3) naturally provides a set of signals at f_{out} where the phases from adjacent ring elements are ideally separated by a fixed angle, so the phase depends on where the output is extracted from the ring, and this can be selected, as illustrated conceptually in Figs. 5.20 and 5.21a. In the latter, when a swallow input advances the number r in the mod-M accumulator by 1, the address sent to the phase-select multiplexer (mux) advances by 1 and selects the next (later) phase from the VCO, introducing a phase change of $1/M$ cycle. This is ideally equivalent to the circuit in Fig. 5.21b, where the VCO operates at a frequency Mf_{out}. The phase-select multiplexer there becomes part of a standard variable modulus prescaler within the

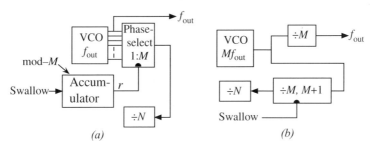

FIGURE 5.21 Oscillator with phase switching (a) and equivalent higher frequency oscillator and variable divider (b).

overall divider, and the $\div N$ input sees a time change of cycle/(Mf_{out}), that is, a phase change of $1/M$ cycle, when a pulse is swallowed. The resulting S_φ is thus reduced by 20 dB $\log_{10}M$, compared to that when a whole cycle is swallowed.

This reduction in quantization noise is, of course, welcome, but may or may not be important in a particular design that is influenced by other noise sources (e.g., Section 4.3).

When a swallow occurs in Fig. 5.21, the period into the $\div N$ circuit is

$$T = T_{\text{out}}\frac{M+1}{M} = T_{\text{out}}(1+1/M), \qquad (5.4)$$

so the period has been lengthened by a fraction $1/M$ of a cycle and the corresponding frequency is

$$f = \frac{f_{\text{out}}}{1+1/M}. \qquad (5.5)$$

If the output frequency is required to be

$$f_{\text{out}} = [N_{\text{int}} + n_{\text{fract}}]F_{\text{ref}}, \qquad (5.6)$$

then the frequency of the equivalent oscillator in Fig. 5.21b must be

$$Mf_{\text{out}} = [MN_{\text{int}} + Mn_{\text{fract}}]F_{\text{ref}}. \qquad (5.7)$$

For example, in a fractional-N (first-order $\Sigma\Delta$) synthesizer, if $n_{\text{fract}} = 1/32$, a cycle would usually be swallowed every $32T_{\text{ref}}$. But, if $M = 4$, $Mn_{\text{fract}} = 4/32 = 1/8$, so a cycle would be swallowed every $8T_{\text{ref}}$ in Fig. 5.21b. The equivalent VCO would see the effect of a 1-cycle swallow at a frequency of $f_{\text{ref}}/8$, producing sidebands at $\pm f_{\text{ref}}/8$. These sidebands are due to phase modulation at a modulation frequency of $f_{\text{ref}}/8$. The magnitude of the modulation will be reduced, due to frequency division, by $M = 4$ at f_{out}, but the modulation frequency will not change. Therefore, the signal at f_{out} will have sidebands due to a 1/4-cycle phase change offset by $\pm f_{\text{ref}}/8$ rather than sidebands due to a 1-cycle change at offsets of $\pm f_{\text{ref}}/32$. Not only will the sidebands be lower by 20 dB $\log_{10}(4) = 12$ dB but also the modulation frequency is four times higher and, therefore, easier to filter.

These are points made by Boon et al. [2005]. They developed a circuit to sequence the phases into the divider by plus or minus one step (or no step) each reference period. ($\div M$, $\div M + 1 \Rightarrow \div M$, $\div M \pm 1$ in Fig. 5.21b.) One problem inherent in this process, which they address, is illustrated in Fig. 5.22. Since the phase switching occurs on the rising edge of the output waveform, it must be delayed until Φ_2 and Φ_1 have the same value or a glitch occurs. Note that a more advantageous switching point was chosen in Fig. 5.20, which, however, does not consider the practical problem of obtaining the waveform to cause switching at that point. Any fixed delay that solves this problem is likely to limit the frequency range of the circuit. Moreover, the problem would

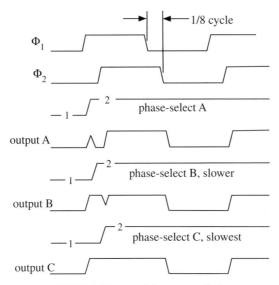

FIGURE 5.22   Possible output glitches.

become worse if we were to change phase by more than one step at a time. (There may also be a problem due to asynchronous switching of mux addresses, such as is addressed in Section A.1.) Note also that if $n_{fract} = 3/4$ in this example, $Mn_{fract} = 3$, so three cycles would be swallowed in Fig. 5.21b every reference period. This would effectively be just an increase in divide ratio, ideally producing no sidebands.

Others have used the configuration in Fig. 5.23 to avoid the glitch and to allow freedom in choosing the number of phase increments to change at one time. Here the various phases are used to sequentially trigger delayed versions of the divider output and the phase-select mux chooses one of these. But now we must account for rollovers, instances when the value r in the mod-M accumulator would exceed its capacity and, hence, roll over to a lower number. In that case the phase from the mux will be reduced and the carry output of the accumulator will be used to insert a whole

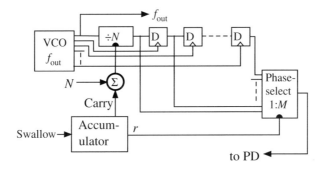

FIGURE 5.23   Phase shifting at the divider output.

cycle into the $\div N$ circuit (i.e., increase N by 1), causing a net phase change of the desired amount.

Lee et al. [2001a] used a 12-phase VCO to generate effective divide ratios of $4P + S + k/12$, where S is the number of swallows in a pulse swallower and k is the number of fractional swallows. Park et al. [2001] used an eight-phase VCO to achieve a divide ratio of $P + k/8$.

5.7.1 Resulting Spurs

A significant difference between Fig 5.21a and b is the tendency for the a configuration to produce spurs as a result of delay inequalities between the multiphase outputs. Any deviation from equal phase increments, between the outputs, leads to distortion of the modulation at the PD. Lee et al. [2001a] and Park et al. [2001] concentrate on shifting the frequency by F_{ref}/M, implying a step of one phase increment each reference period. (This is equivalent to a change of 1 in the overall divide ratio of a synthesizer using the configuration in Fig. 5.21b, a change that should produce no sideband in that configuration.) Distortion caused by unequal phase increments produces sidebands that are offset by multiples of the frequency shift. Park et al. attacked this problem with a calibration loop that adjusts the delays toward equality, reducing the first modulation sideband from -33 to -55 dBc.

5.7.2 Estimate of Achievable Suppression

The -55 dBc sidebands, achieved by Park et al. [2001], equal 20dB $\log_{10}(m/2)$, so $m|_{dB} = -49$ dBr and the rms phase deviation is $\sigma_{dB} = -52$ dBr. If this were spread over $F_{ref}/2$ (Appendix N), the density would be

$$\frac{\sigma^2}{10^7 \text{ Hz}} \Rightarrow -52 \text{ dB rad} - 70 \text{ dB Hz} = -122 \text{ dBr/Hz} \Rightarrow -125 \text{ dBc/Hz} \qquad (5.8)$$

in their case, where F_{ref} was 20 MHz. This is about 14 dB below the level of $\mathcal{L}_{flat}$ that we used in Fig. 4.11 to represent an insignificant level of noise due to circuit defects. While there may be some phase noise in other sidebands also, this still suggests that it may be possible to overcome the phase mismatch problem at the cost of increased complexity.

5.7.3 Fractional Swallows in a ΣΔ Synthesizer

Encouraged by similar computations, Riley and Kostamovaara [2003] used ΣΔ modulation in a Simulink simulation to spread the energy of the F_{ref}/M spur and showed that the spurs due to phase mismatch could be suppressed by passing LSBs of n_{fract} through a ΣΔ modulator. They used 16 phases and controlled the phase-select mux with the 4 MSBs of n_{fract}, passing the other bits through high-order ΣΔ modulators. A fifth-order modulator suppressed -55 dBc spurs at $\pm f_{fract}$ to -60 dBc, while providing much greater suppression to sidebands offset by multiples of f_{fract}. When the modulator order was changed from 5 to 6, the suppression of the first sidebands increased 14 dB. The ability of the modulation to suppress the sidebands is welcome, but it is unfortunate that its order had to be so high. Some related work suggests that it might be worthwhile to try adding uniformly distributed numbers rather than $\Sigma\Delta$ modulation (Section 9.1.4).

This study also illustrates some of the complexities involved in employing fractional swallows with $\Sigma\Delta$ modulation, including the sharing of control, and overflow, between the circuits that determine N and the phase selection. The LSBs of n_{fract} are used to inject the corresponding value into the (average) synthesized frequency, as in a $\Sigma\Delta$ synthesizer, but the MSBs are not. Their value is more like N in an integer-N synthesizer; it determines the fractional phase shift that occurs each reference period and which would cause no sidebands if the phases were perfect (as in the example above where n_{fract} was 1/4). Since we here depend on $\Sigma\Delta$ processing of the LSBs to randomize the phases in order to suppress the sidebands caused by their imperfections, we will lose that randomization at frequencies where the LSBs of n_{fract} are zero. We have seen that one technique for avoiding discrete sidebands is to avoid certain frequencies and that this does not cause a problem if we have enough bits. This may just increase the number of frequencies to be avoided.

5.8 HARDWARE REDUCTION

It may be possible to save hardware by reducing the length of accumulators after the first, dropping some of the less significant bits between the accumulators. Ye and Kennedy [2009a] have developed a design procedure for reducing hardware complexity in $\Sigma\Delta$ modulators without sacrificing performance in MASH and in single-quantizer modulators [Ye and Kennedy, 2009b]. They report a typical 20% reduction in circuit size of a third-order MASH modulator.

5.8.1 Analysis

The length of an accumulator is truncated by dropping n_{trunc} least significant bits. This process occurs in the MASH structure of Fig. 5.24 at one (or more) of the r outputs, and it is commonly modeled as noise injected at that point. The efficacy of that model depends on lack of correlation between the noise sources in the model.

Let us write the length of the mth-accumulator (A_m in Fig. 5.24) as $2^{n(m)}$. Then, the truncation noise source consists of the numbers $[0, 1, 2, \ldots]2^{-n(1)}$ up to, but not including, $2^{-n(m)}$, which are discarded at its input. We can write the discarded numbers as[37]

$$[0, 1, \ldots, (2^{n(1)-n(m)}-1)]2^{-n(1)}. \tag{5.9}$$

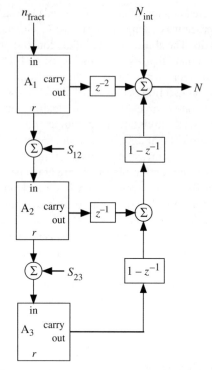

FIGURE 5.24   MASH-111 with truncation noise.

If the numbers are equiprobable, the variance of this sequence is

$$\sigma_{Tm}^2 = \frac{2^{-2n(m)} - 2^{-2n(1)}}{12}. \tag{5.10}$$

When a large number of bits are dropped [i.e., $n(m) \ll n(1)$],[38]

$$\sigma_{Tm}^2 \approx 2^{-2n(m)}/12. \tag{5.11}$$

A noise source introduced at the input to an accumulator passes to the output ("carry out"), multiplied by the accumulator's forward transfer function z^{-1} (for the model assumed in Fig. 5.24). If we can model the dropped number sequence as being random, a white noise source is produced with two-sided density:

$$S_{2,(m-1)m} = \frac{\sigma_{Tm}^2}{F_{\text{ref}}}. \tag{5.12}$$

The white character of this noise source (spur power independent of spur frequency) is assumed by Ye and Kennedy and also by Hill [1997], on whose work

Section F.8.3.5.5 is based but what we learned in Section 2.1.4 should cause us to be suspicious of this assumption.

The noise density $S_{2,(m-1)m}$ induces a SSB PSD, at the synthesizer output, of

$$\mathcal{L}_{Tm} = S_{2,(m-1)m}(2\pi)^2 \left[2\sin\left(\frac{\pi f_m}{F_{\text{ref}}}\right) \right]^{2(m-2)}. \tag{5.13}$$

We obtain this relationship by following the same process that produced Eq. (2.3), since this truncation noise and the quantization noise are processed similarly. Truncation at the input to the second accumulator passes through the accumulator with just a delay and is multiplied by $(1 - z^{-1})$ in the MASH structure. It is then divided by $(1 - z^{-1})$ when being translated to phase, so the resulting $\mathcal{L}_{Tm}$ *is not* a function of f_m. Truncation at the input to the third accumulator produces PSD that *is* a function of f_m for similar reasons.

The deleted numbers consist of n_{trunc} LSBs of the original sequence, where

$$n_{\text{trunc}} = n(1) - n(m) \tag{5.14}$$

and $n(1)$ is the number of bits in the first accumulator (and in all the accumulators before truncation). Since they were processed as the LSBs in the preceding (not truncated) accumulator, they behave as if they were processed in an accumulator with n_{trunc} bits. If n_{fract} is odd, which we will assume from here, their sequence length is $2^{n_{\text{trunc}}}$, which may be short enough to make the discrete nature of the spurs significant (because they will be relatively widely separated). The spur separation and the offset of the first spur from spectral center will be

$$f_1 = F_{\text{ref}} 2^{-n_{\text{trunc}}}. \tag{5.15}$$

We obtain the relative amplitude of the spurs by multiplying the PSD by the spur separation,

$$
\begin{aligned}
\mathcal{L}_k &= f_1 \mathcal{L}_{Tm}(k f_1) \\
&= \sigma_{Tm}^2 (2\pi)^2 2^{-n_{\text{trunc}}} \left[2\sin\left(\frac{\pi f_k}{F_{\text{ref}}}\right) \right]^{2(m-2)}.
\end{aligned} \tag{5.16}
$$

If sufficient bits are dropped for Eq. (5.11) to hold, this is

$$\mathcal{L}_k \approx \frac{(2\pi)^2}{12} 2^{-[2n(m) + n_{\text{trunc}}]} \left[2\sin\left(\frac{\pi f_k}{F_{\text{ref}}}\right) \right]^{2(m-2)}. \tag{5.17}$$

For the second accumulator,

$$\mathcal{L}_k|_{\mathrm{dB}} \approx 5.2 \text{ dBc} - 6 \text{ dB}n(m) - 3 \text{ dB}n_{\mathrm{trunc}} \qquad (5.18)$$

$$= 5.2 \text{ dBc} - 6 \text{ dB}n(1) + 3 \text{ dB}n_{\mathrm{trunc}}. \qquad (5.19)$$

Ye and Kennedy have made use of the relatively large separation between spurs to create a design strategy where the truncation spurs have amplitude lower than that of the quantization noise spurs at the same frequency. The quantization noise, Eq. (2.3), has a steeper low-frequency slope than does the truncation noise of Eq. (5.13). Therefore, below some value of f_k, the truncation noise envelope will exceed the quantization noise envelope. By designing the truncation scheme such that f_1 is above that frequency, we can ensure that the truncation noise is covered by the quantization noise (i.e., no truncation spurs will exist below the frequency where they would be larger than the quantization spurs). In applications where density is more important, this solution would be pessimistic, since the spur density is greater in the quantization noise.

While making the truncation spurs lower than the quantization spurs is important for showing that hardware reduction is feasible, the ability to compute what spurs are produced by truncation is important in system design.

Note that $n(1)$ rather than $n(m-1)$ appears in Eqs. (5.10) and (5.14). The truncation is theoretically accounted for by adding noise sources to the configuration where all accumulators still have $n(1)$ bits, so a truncation of the third accumulator is considered a reduction in bits relative to this, even if the second accumulator has been truncated. If length is reduced only between the first and second accumulators, there is still a noise source at the input to the third accumulator, because it has reduced bits relative to $n(1)$. However, if there is no additional truncation there, the noise level at the synthesizer output due to truncation of the third accumulator will be lower than the noise caused by truncation of the second accumulator or, at higher f_k, by the quantization noise.

5.8.2 Simulation

Figure 5.25 shows how truncation can be introduced into a Simulink model. The word from accumulator $m-1$ is multiplied by $2^{n(m)}$. This causes bits with value $2^{-n(m)}$ or greater to become integers, while lesser bits remain fractional. The "floor" function rounds numbers to the next lower integer, deleting the fractional bits. The final amplifier restores the number to its original value, less bits lower than 2^{-m}. We can monitor the process by viewing the difference between the input and output of the truncation circuit on a scope. At the first truncation, this waveform is the injected noise. For subsequent truncations (higher m), it generally differs from the theoretical noise source, which is the difference between the truncated input and what would have existed if there had been no preceding truncation.

One of the first things we would be inclined to do with such a model is to truncate the input to the second accumulator, say to 8 bits (from 16), and see how our previous

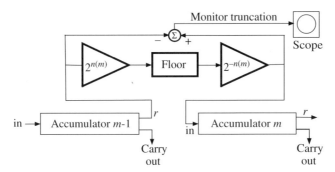

FIGURE 5.25   Simulating truncation

model of a MASH-111 synthesizer, with $n_{\text{fract}} = 0.0625$ and a 2^{-16} seed, responds. We may look for perturbations of the spectrum in Fig. 2.31 and would be surprised to observe a spectrum that looks like Fig. 2.27 instead. The randomizing effect of the seed requires it to be passed from the first accumulator to subsequent stages; truncation at the input to the second accumulator prevents this. For this reason, we depend on an odd n_{fract} in the development above.

Next we may add an LSB to n_{fract} in order to randomize the sequence, setting the input to $(n_{\text{fract}})_A = [0.0625 + 2^{-16}]$. According to Eqs. (5.15) and (5.19), with $n_{\text{trunc}} = 8$, we should see sidebands offset by multiples of $f_1 = 39.0625$ kHz with relative amplitudes of about -67 dBc plus $H(kf_1)$ [we will subtract $H(kf_1)$ in all the following], but we observe a level at f_1 that is 17 dB higher, -50 dBc. We again see this same spectrum if we set $(n_{\text{fract}})_B = [0.0625 + 2^8 - 2^{-16}]$. However, when the input is $(n_{\text{fract}})_C = [0.0625 + 15 \times 2^{-16}]$, we see a first sideband drop 24 dB to -74 dBc, 7 dB below the value given by Eq.(5.19).

It would appear that S_{12} is not white when n_{fract} equals $(n_{\text{fract}})_A$ or $(n_{\text{fract}})_B$. At these values (and others offset by 2^{-8} from them), the input to the lower 8 bits consists of just the LSB, resulting in a 39.0625 kHz sawtooth (refer to Section 2.1.4) with peak-to-peak amplitude of $2^{-n(m)}$. The fundamental Fourier component of this sawtooth has amplitude $2^{-n(m)}/\pi$. We multiply by 2π rad to get the peak phase deviation m and divide by 2 (for $m \ll 1$, Section F.3.1) to get the relative sideband amplitude, $m/2 = 2^{-n(m)}$, in decibels, $-6n(m)$ dB. For $n(m) = 8$, this is -48 dB, approximately (2 dB above) the level that we observed. In general, the power in this component exceeds the power given by Eq. (5.18) by $[3n_{\text{trunc}} - 5.2]$ dB.

When more of the lowest 8 bits were set in $(n_{\text{fract}})_C$, the first sideband was close to what would be expected under the white noise assumption. We might also expect that assumption to apply to S_{23}, even with inputs like $(n_{\text{fract}})_A$, since the deleted bits were processed by two accumulators in series. However, the first sideband due to S_{23} appears to be about 20 dB larger with $(n_{\text{fract}})_A$ than with $(n_{\text{fract}})_C$ [which is about 6 dB below the level given by Eq. (5.17)].

We may consider employing dither (Section 4.2.1) to whiten the sequence, but this would seem to require further experimentation. Unshaped dither ($q = 0$) of sufficient magnitude appears to suppress the spurs but adds low-frequency noise. Shaped dither

(17th LSB, $q = 1$) is ineffective in reducing the spurs. Moreover, when the seed is reset ($n_{\text{fract}} \Rightarrow 0.0625$), shaped dither in the 17th LSB does not prevent the discrete spurs, as in Fig. 2.27, from reappearing, reminiscent of the effect that the truncation had on the seed.

Without further analyzing the applicability of the white noise assumption, let us take what we have found as an illustration of the value of simulation (while still appreciating the value of the theory to our understanding of the effects of truncation).

CHAPTER 6

SIMULATION

Much of our discussion to this point has been based on Simulink® models. These have been used to explain fundamental relationships in the $\Sigma\Delta$ synthesizers and to expose performance defects (but without attempting to model all the circuit imperfections and limitations in a real synthesizer). A collection of models is provided online for use in further study and for expansion to represent additional effects and to answer new questions. Here we will discuss these models and some MATLAB® scripts that have been used as well.

We will depend on the Simulink Users Guide and Help documents to explain the details of using the Simulink program to those who are not familiar with it; we will concentrate on the information that is unique to our models and simulations.[39] The models and scripts must be downloaded (Appendix W) and put into a directory that is included in the MATLAB search path (Command Window/File/Set Path...) in order to be accessed by the Simulink or MATLAB program. Required block sets, and toolboxes required for MATLAB scripts, are given in Appendix T.

Our first model description will include many items that are carried over to other models and which will not be discussed again.

6.1 SandH.mdl

The early parts of Chapter 2 use the Simulink model `SandH.mdl` shown in Fig. 6.1.

Advanced Frequency Synthesis by Phase Lock, First Edition. William F. Egan.
© 2011 John Wiley & Sons, Inc. Published 2011 by John Wiley & Sons, Inc.

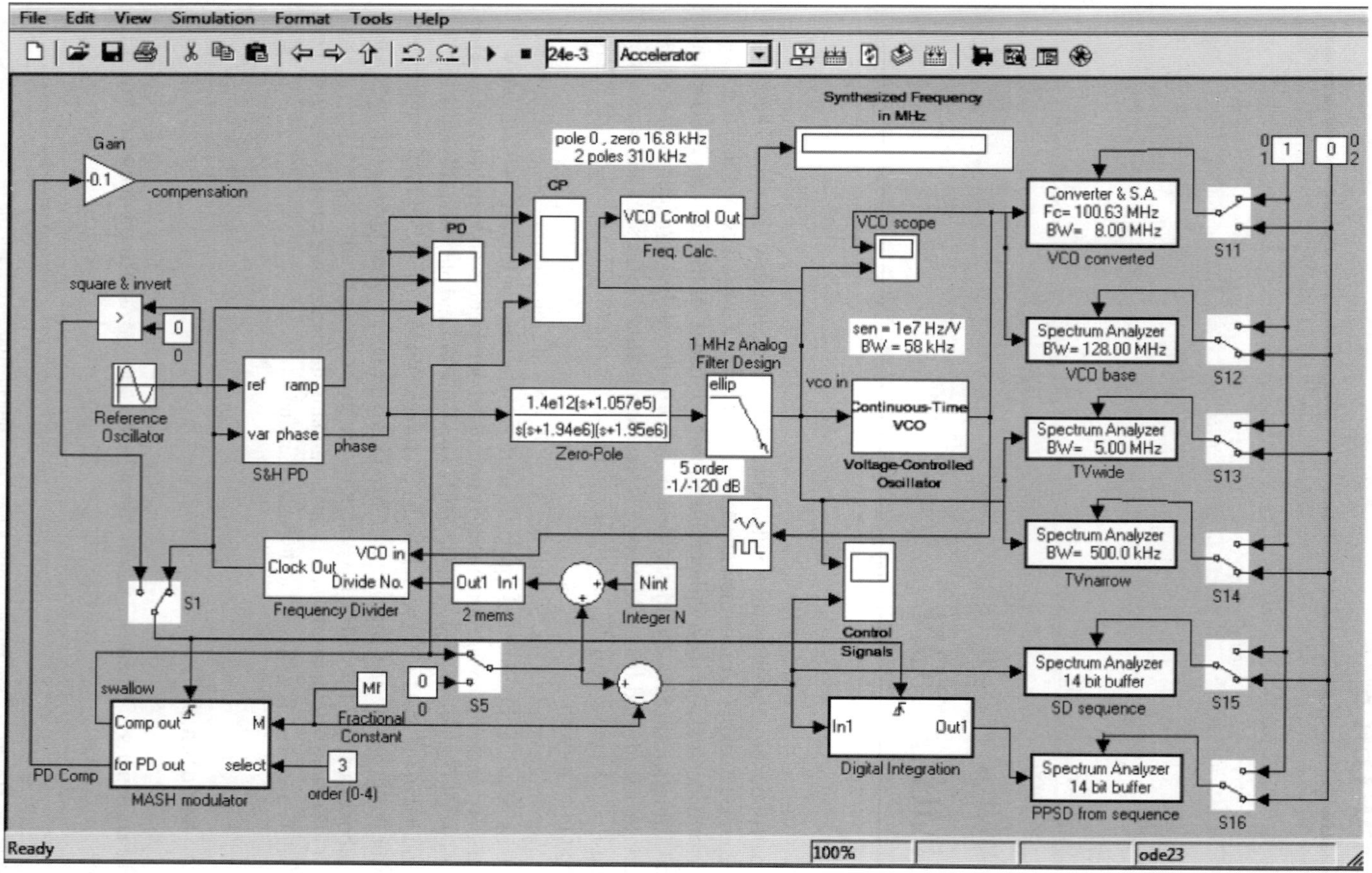

FIGURE 6.1 S&H.mdl.

6.1.1 The Synthesizer Loop

The "Continuous Time VCO," which we will use in all of our models, is one of the oscillators available from the Simulink Library (from model window, pull down View/Library Browser). Its output is a sine wave at frequency f_{out}, which is computed from its tuning input and displayed on the "Synthesizer Frequency" display above it. The output is fed to time domain and frequency domain displays and to the "Frequency Divider." The divider output feeds the S&H PD (sample-and-hold phase detector). The output of the PD passes through the loop filter, consisting of "Zero-Pole" and a fifth-order elliptic filter, into the VCO to control its frequency. This is the synthesizer loop.

6.1.2 MASH Modulator

The MASH $\Sigma\Delta$ modulator is in the lower left. Its order can be set between 0 and 4 and its input n_{fract} is designated Mf. Outputs are the swallow signal to the frequency divider and the analog cancellation signal, PD Comp, from the last accumulator. The swallow signal is added to N_{int} (Nint) and PD Comp is multiplied by a constant, approximately $-1/N$, and sent to the "CP" scope for comparison with the PD output.

Double-click the MASH modulator block to see its diagram. Within that diagram, double-click Accumulator1 to see its model. Within Accumulator1, double-click Unit Delay to see the seed.

6.1.3 Setting Parameters

Certain parameters are set in the Model Workspace using the Model Explorer (View/ Model Explorer or Control-H, and then expand files on the left to find the workspaces). These are shown in Fig. 6.2.

The parameter seed is the initial value set into the first accumulator. Here it is 2^{-16}. The reference oscillator is set to a frequency of fref Hz. Mf is n_{fract} and Nint is N_{int}. The variable sen equals K_v in Hz (Hz/V if we consider the control signal to be in volts). The initial frequency of the VCO is fi Hz. When we are interested in observing the steady-state performance (with $\Sigma\Delta$ modulation), we set fi to equal fref(Nint + Mf) so that the loop settles quickly, although we will usually have to wait for an initial phase error to settle. (Double-clicking on the VCO allows us to set the initial phase.)

The length of binary accumulators in the $\Sigma\Delta$ modulator is determined by the value of the lowest bit used in the simulation. This will be set either by the seed (assuming it to be a single bit) or by the LSB in Mf, whichever is smaller. (If we could somehow simulate a modulator longer than this, the additional length would be devoid of content and would have no function in the model.) The models do not quantize these inputs, as if the number of available bits is unlimited. We could limit the size of the accumulators by preceding the first accumulator by a truncate circuit, such as Truncate2 in Section 6.8, but we depend on the user to use binary fractions for

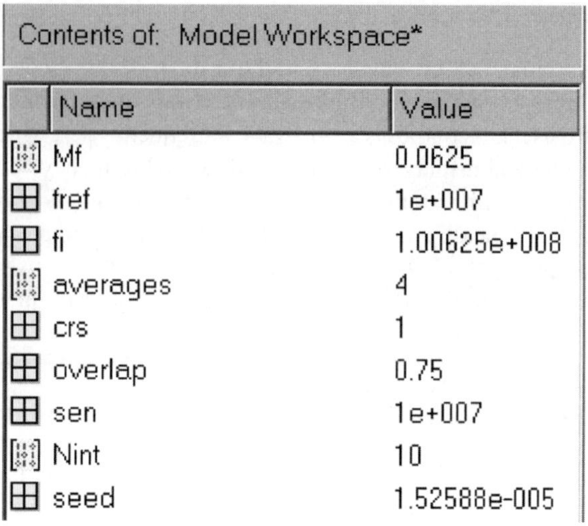

Contents of: Model Workspace*	
Name	**Value**
[##] Mf	0.0625
⊞ fref	1e+007
⊞ fi	1.00625e+008
[##] averages	4
⊞ crs	1
⊞ overlap	0.75
⊞ sen	1e+007
[##] Nint	10
⊞ seed	1.52588e-005

FIGURE 6.2   Model Workspace parameters.

`Mf` and `seed`. The Simulink program allows us to enter parameters as equations; for example, Mf can be entered as "2^-4 + 2^-16."

The rest of the parameters refer to spectrum analysis, which is discussed in Appendix S. A value of 4 for `averages` means that each displayed spectrum in the spectrum analyzers (SAs) is an average of the last four computed. A value of 1 for `crs` means that the noise bandwidths B_n in the spectrums are nominal; they are all multiplied by `crs` (which stands for "coarseness"). An `overlap` of 0.75 means that 75% of the data for each frame of the spectrums is shared with previous frames. That is, a new FFT is computed whenever 25% of the buffer has been filled with new data, so it takes four frames to completely change the data.

Parameters can also be set in the Base Workspace. They then apply to all models, but if a parameter is set in both workspaces, the value in the Model Workspace takes precedence. Therefore, we must delete the parameter in the models to have the value in the Base Workspace be effective.

6.1.4 Accumulator Size

The sequence of numbers in a modeled accumulator is simply related to the seed and n_{fract}, without the necessity of specifying a size (capacity) for the accumulator. When the contents of an accumulator in a MASH modulator reaches 1(which just exceeds its capacity), a carry will be generated and the contents will then be reduced by 1. The required accumulator size would be the least common denominator of the fractions that are entered into it. The binary case is discussed above, but, in general, other bases can be modeled by entering the appropriate fractions. If $n_{\text{fract}} = N/509$, the simulated accumulator must have length 509; it would take a total of 509 single-bit inputs to increase the contents from 0 to 1. But, if a seed should be set, the simulated accumulator would have to accommodate it also. If the seed should be 2^{-10}, for example, the accumulator would have length 521,216 [$2^{-10} = 509/521,216$ and $N/509 = 1024N/521,216$]. Other accumulators that are fed the contents of a previous accumulator must be as long as the previous accumulator, so they can accommodate the same numbers. In most cases, we do not have to model the specific implementation that causes the accumulator to have a particular base.

6.1.5 Scopes

Double-click a scope to open it. Right-click in a scope display and choose "Axes Properties ..." to adjust the y-axis range. Click the "parameters" icon above the scope displays to change the horizontal axis or the sample rate (space between plotted points—you cannot change this while the simulation is running). Figure 2.3 shows a CP Scope display. Scopes can be easily inserted or deleted and connected to various signals.

6.1.6 Spectrum Analyzers

If you double-click a spectrum analyzer, you get a dialog in which you can set the display bandwidth or number of bits, whichever parameter is displayed on the analyzers icon. If you right-click the SA and select "Look under mask," you can see the underlying diagram, including the Spectrum Scope. If you then double-click the Spectrum Scope, you will see a window with several tabs where you can specify properties. Some of these properties are written in terms of the variables in the Model Workspace. If you choose the "Axis Properties" tab, you can select the y-axis display range. You can also set the "Frequency display limits" to "User-defined" if you want to view only a portion of the band; then you can select what portion. In the "Display Properties" tab, you can check to open the scope (display) at the start of simulation or you can open it immediately (which, however, seems to delay the start of frame 1 until a whole number of frame periods have gone by[40]). You can also set the SA's display position there, which might be especially important if it is off screen.[41] You can also set the display window to its current position by choosing Axes/Save Scope Position from the menu above its display.

Each spectrum analyzers can be enabled[42] by connecting the corresponding switch to 1. Disabling a spectrum analyzer increases the simulation speed by about half as much as deleting that spectrum analyzer from the model, which you can also do. If a model is to be used extensively, it would be efficient to run a copy of that model where unneeded analyzers have been deleted.

With a 0.75 overlap, it takes four frames before the buffer is full. When that has occurred, the frame number displayed will be 5. You may have to wait longer, maybe about twice as long, until the loop has settled enough to show the steady-state spectrum.

The simulation time for a frame depends on the sample rate, which is tied to the specified bandwidth of the analyzer, and the size of the buffer (Appendix S). Figure 6.3 shows the spreadsheet SpectrumAnalyzers.xls, which can be used to compute spectrum analyzer parameters. It also gives an estimate of the corresponding

Enter ⇓   ⇓ Use Tools/Protection/Unprotect Sheet to change buffer size.
Loop BW 58000 Hz   ⇒ Time Constant τ ≈   2.74 μsec

crs   1

	BW (MHz)	buffer	sample rate (MHz)	analysis BW (Hz)	simulation time/frame (μsec)	appx. execution time /frame (minutes) accelerated	normal	τ's per frame
VCO Coverted	8	8192	20.48	3785	100	0.7	2.6	36
VCO base	128	65536	327.68	7570	50	0.4	1.3	18
TV wide	5	4096	12.8	4731	80	0.6	2.1	29
TV narrow	0.5	512	1.28	3785	100	0.7	2.6	36
	bits			Enter ⇒	300	2.2	7.7	
SD Sequence	14	16384	10	924	410	3.0	10.5	149
PPSD from Sequence	14	16384	10	924	410	3.0	10.5	149

FIGURE 6.3 SpectrumAnalyzer.xls spreadsheet. Execution times are approximate and vary with the model, what is active in the model, and the computer processor.

execution time, based on data that the user enters in the box. (The user obtains those numbers by timing a simulation.) Note how much time would be required to get 8 or 10 frames of, for example, "SD Sequence" (i.e., $\Sigma\Delta$ sequence), especially in the Simulink "normal" mode. The accelerated mode (see top of window in Fig. 6.1) is about 3.5 times faster.

We have observed something close to the final spectral shape occurring in the eighth frame of the "VCO Converted" SA. By that time the loop has had many time constants to settle. See the last column in Fig. 6.3. While the loop pull-in is too complex to be characterized by a single time constant, the reciprocal of the loop bandwidth (upper right) is a reasonable approximation for our purposes here. Long frame times can be shortened by increasing crs at the cost of a coarser spectrum— and conversely.

6.1.7 Spectrums Observed

The top SA converts the signal from the VCO, at f_{out}, to a frequency equal to half of the SA's specified bandwidth, as shown in Fig. 2.7 for example. In this case, the frequency is 4 MHz. To do this, the signal at $F_{out} = 100.625$ MHz was mixed with a fixed signal at 96.625 MHz (Fig. 6.4). The 1 MHz elliptic filter in the *loop* is

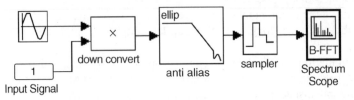

FIGURE 6.4 Frequency converter in SA.

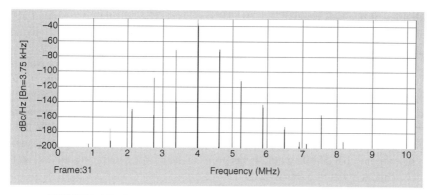

FIGURE 6.5 SA display corresponding to Fig. 2.7 (at a later frame).

important for a proper spectral display. It must attenuate spurs that are more than 4 MHz from the center of the spectrum to keep them from folding over on top of the observed spectrum (Section S.8.2). For example, a spur 5 MHz below F_{out} at 95.625 MHz would mix with the fixed 96.625 MHz and appear at 1 MHz in Fig. 6.5. The low-pass elliptic filter in the SA would not attenuate such a signal. However, that filter is important in attenuating other spurs that can distort our spectral picture.

Figure 6.5 shows the whole display that appears on the SA, the one from which Fig. 2.7 was derived. Note the signal at 7.55 MHz and two sidebands 625 kHz either side of it. These were not shown in Fig. 2.7 to prevent confusions, but we now want to understand their origin. The signal at 7.55 MHz is a frequency-translated version of the main response at 4 MHz, attenuated 120 dB by the filter in the SA. It is the sum frequency (the image) from the mixing process, 96.625 MHz + 100.625 MHz = 197.25 MHz, translated by 10 times the sampling frequency (Section S.8.3). The sampling frequency is 2.56 times the specified bandwidth, 8 MHz, or 20.48 MHz. So, −197.25 MHz (we need to use both positive and negative frequencies in this process) is translated by 204.8 MHz to 7.55 MHz, which we see it in Fig. 6.5. Here, the loop filter is not effective, since the signal originates from the desired signal, but the filter in the SA is. The attenuation is just insufficient to keep us from seeing the spur on this display. (+ 197.25 MHz is also translated to −7.55 MHz.)

Looking for anomalous spurs by turning off the fractional-N modulation, we find the highest in the Converter & SA display to be 170 dB below the signal. This would be just below the visible area in Fig. 6.5. There also appears to be some close-in noise at about −200 dbc/Hz, but that is reduced significantly when the modulation is on. Noise levels on the tuning line ("TVnarrow" and "TVwide" SAs) correspond approximately to the noise observed on the output signal.

The second (from top) SA shows the unconverted synthesizer output. Both this and the top SA display relative single-sideband PSD, L.

The third and fourth SAs show the spectrum of the tuning voltage at two different display bandwidths. They are calibrated in frequency power spectral density (FPSD) with units of decibels relative to a Hz^2/Hz (i.e., Hz). They display two-sided FPSD, which when divided by f_m^2 gives the two-sided PPSD, $S_{2\varphi}(f_m) = \mathcal{L}_\varphi(\Delta f = f_m)$. In most cases, this equals $L(\Delta f)$, as displayed on the upper two SAs.

The fifth SA, labeled "SD Sequence" ($\Sigma\Delta$ sequence), displays the output from the $\Sigma\Delta$ modulator. The last SA, labeled "PPSD from sequence," displays that output, after digital integration, in units of phase density, indicating the PPSD that it theoretically produced as the synthesizer output less the filtering effect of the loop [i.e., before multiplication by $H(f_m)$]. There are no filters in these two analyzers and they sample at the same rate as the $\Sigma\Delta$ modulator's clock, F_{ref}, so they analyze the digital streams. Ideally, the number of bits would be as great as the number in the modulator, so the entire sequence would be analyzed.

6.1.8 Reason for Frequency Conversion

Frequency conversion is commonly used in analog SAs to enable the spectrum to be examined closely in the region of interest, and it was originally used for that purpose in these models. However, with the improvement in the Spectrum Scope model that allows its display to show any portion of its maximum frequency range ("Frequency display limits: User-defined"), this advantage has been reduced considerably.

For example, if the bandwidth BW in the upper "Converter and SA" is set to 8 MHz, the sampling rate will be 2.56 times higher, 20.48 MHz. If the buffer is set to 8192 bits, the bin width will be [20.48 MHz/8192 =] 2.5 kHz and the time to fill the buffer will be [8192 c/20.48 MHz =] 400 μs. If we now increase the bandwidth 16 times to 128 MHz, as in the second "VCO base" SA, the bin width (and corresponding noise bandwidth) will increase by 16 times, but the time to fill the buffer will decrease by the same ratio. Therefore, we can now increase the buffer size by 16 times and reestablish the original bin width and time. With the ability to zoom in on the desired frequency range, we can obtain the same frequency resolution in the same frame time without frequency conversion as we can with frequency conversion. Moreover, we can avoid the spectral folding that occurs in the frequency conversion and which places restrictions on our synthesizer's spectral width. And, we have a better chance of making T_{segm} larger compared to T_{sequ}, and, thus, obtaining something like Fig. 2.15 rather than Fig. 2.14 (Section S.1).

One advantage that frequency conversion still has, however, is that it enables us to control the center frequency of our display. With 5 kHz bins, for example, the synthesized frequency must be a multiple of 5 kHz to center its spectrum in a bin and avoid spectral leakage (see Section S.3). With frequency conversion, the upper spectrum analyzer can be set to automatically center on the signal, accounting for the effect of N_{int} and n_{fract} and even `alpha`, which modified the fraction in the case of the `HandK.mdl`. The signal is centered at BW/2 and the bin size is 2.56 × BW/L_{buf}, where L_{buf} is the length of the buffer. Therefore, the number of bins from 0 to the center frequency is $L_{buf}/5.12$. The spectrum will be centered in a bin if this is a whole number, requiring that L_{buf} (assuming it is binary) be at least $2^7 = 128$. Typically we use 2^{13}.

6.1.9 Synchronization

The switch S1 controls whether the ΣΔ modulator is synchronized to the reference signal or to the divider output. The former provides a constant clock frequency, whereas the latter, which is the usual configuration, gives a phase-modulated clock. The switch provides some experimental flexibility but it does not affect the output spectrum because the divider does not read the carry output until a count sequence begins. The only effect observed due to changing synchronization was on the spectrum at the output of the Digital Integration block when it was observed as an analog signal. Then the modulated clock produced additional noise, perhaps foreshadowing the effect on the output spectrum of varying sampling rate (Section 2.3).

6.2 SandHreverse.mdl

The inputs to the S&H PD can be reversed (Section 2.3.1) with a few changes, which are implemented in this model. Double-click the PD block to see the changes. The initial condition of the integrator is also changed, which can be observed by double-clicking the "Integrator" block therein.

6.3 CPandI.mdl

We replace the S&H PD with a PFD and charge pump in CPandI.mdl (Fig. 6.6). The block "Ideal C. Pump & Integrator" contains both the charge pump and the first element in the loop filter, an integrator (double-click the block to see). Following this is a S&H circuit synchronized at half a reference cycle after the reference instant. This enables the PD output, after integration, to change at a constant sample rate if switch S3 is so positioned; otherwise, the S&H is bypassed. This model also adds displays of several important variables, so we can better know the conditions under which the simulation is being run and record them by recording the picture of the model. These variables are crs, which adjusts the coarseness of the SA displays; seed, the initial value in the lowest register of the modulator; and Mf ($= n_{\text{fract}}$), the fractional number. These refinements can be copied to other models with relative ease. This model is used in Sections 2.3.3 through the end of Chapter 2.

6.4 Dither.mdl

Dither.mdl provides LSB dither (Fig. 6.7), as described in Section 3.1. The "Random Integer Generator" is set to generate evenly distributed random integers of values 0 and 1 (double-click the generator to see the dialog). When both S4 and S6 are connected to 0, the random number is multiplied by the LSB value, here 2^{-10}. This value can easily be changed by double-clicking the gain block and altering the value of the gain. The resulting number is added to Mf (n_{fract}). When we simulate a

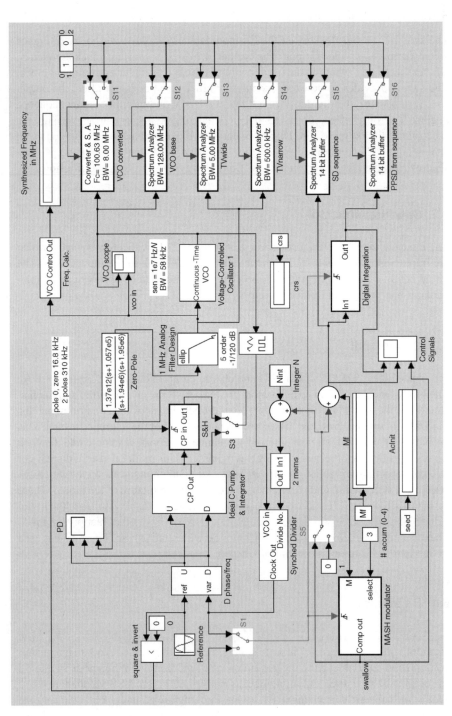

FIGURE 6.6 CPandI.mdl.

112

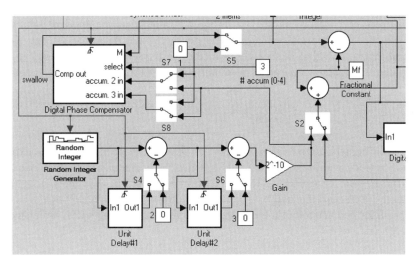

FIGURE 6.7   Dither generator.

simple addition of jitter by toggling of the LSB, `Mf` should not contain a bit as small as the LSB.

Each of the switches S4 and S5 inserts a factor $(1 - z^{-1})$ in series with the random generator when connected to a Unit Delay. When the jitter is sent through S2 to the modulator input, the number of switches so connected equals q in Eq. (3.1). Other switches allow the jitter to be inserted at the input to the second or third accumulator, increasing q by 1 or 2, respectively.

6.5   HandK.mdl

An input has been added to the MASH modulator (Fig. 6.8) in `HandK.mdl` that allows the effective length of the accumulators to be shortened by a, as in Fig. 3.7, for simulation of HK-MASH (Section 3.2). The variable `alpha` is a normalized to the length of the accumulator (since the capacity of the accumulators in the models is normalized to 1); `alpha` $= a/2^n$, where n is the number of bits in each accumulator.

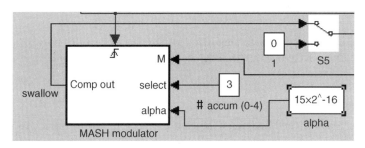

FIGURE 6.8   Modulator modification in HandK.mdl.

We can type "alpha" in the box in Fig. 6.8 rather than a numerical value and then specify the numerical value of alpha in the Model Workspace (Control-H). A value of alpha that is so entered in the workspace will be included in the computed frequency for purposes of centering the spectrum in the upper spectrum analyzer display. This resulted in the centered display in Fig. 3.8c, whereas the offset in Fig. 3.8b was produced by entering the value of alpha directly and leaving it at 0 in the workspace.

The workspace includes a seed to specify an initial value in the first accumulator, as in other models. Initial values can also be inserted in other registers for experimentation (look within the 1/z unit delays within the accumulators in the model) by specifying alseed (alseed was set to 0 in Section 3.2). Variables seed and alseed are also normalized.

The simplified version of the HK-MASH modulators in Fig. 3.12 is simulated in HandKsimple.mdl.

6.6 SimplePD.mdl

SimplePD.mdl employs a simple approximation to the CP, wherein the current source is replaced by a resistor that is switched to 5 V or ground or open (Fig. 4.14). Its weakness is that it requires 2.5 V on the capacitor for the proper currents to be generated. This would occur if 2.5 V produces the synthesized frequency, but it varies with tuning and dynamically with short-term variations in the capacitor voltage (e.g., due to the $\Sigma\Delta$ modulation), effects that are explored in Section 4.2.5. By double-clicking the "RC Charge P" block, we can see the components represented in Fig. 4.14, where $K_p = 4.8$ mA/cycle (= 2.5 V/520 Ω c; double-click on the switch to see that it has 10 Ω internal resistance). This is multiplied by the impedance of the integrating capacitor, $1/[KLF \times 2.2 \times 10^{-7}F \times s]$, which is part of the loop filter. The factor KLF is canceled by the same factor in the remainder of the loop filter. By changing KLF in the Model Workspace (Control-H), we can change the value of the capacitor, as in Fig. 4.15, while maintaining constant loop gain.

6.7 CPandIplus.mdl

Additional enhancements have been added to CPandI.mdl in CPandIplus.mdl (Fig. 6.9). These have been used in the study of defects in the synthesizer circuitry.

6.7.1 CP Balance

Above the "Ideal CP and Integrator" block is a block labeled "balance." The value in this block (double-click to change) multiplies the up charge pump current, producing a mismatch between up and down currents. It is used in Section 4.2.3.

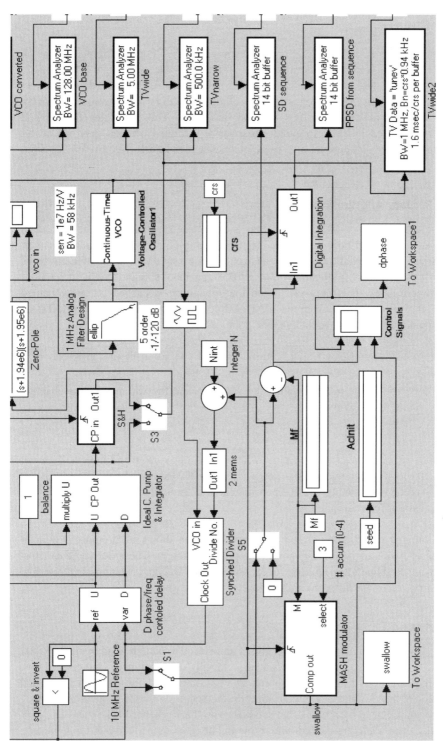

FIGURE 6.9 Additional features in CPandIplus.mdl.

115

6.7.2 PFD Delays

Another controllable imperfection is hidden within the "D phase/frequency" block to the left. Within that block are a pair of "Transport Delay" blocks. The delay in each of these blocks can be set separately. They are set to 0 to cause the PFD to act like the PFD in other models. If they are both set to 1 ns, for example, opposing (canceling) 1 ns current pulses will be generated. This simulates the method commonly used in PFDs to avoid crossover distortion. However, they can be set to different delays, producing a delay mismatch, whose effects on reference sidebands are described in Appendix R.

6.7.3 Data Acquisition

The blocks swallow and dphase can be used to acquire data during the simulation and store it under their respective names. Double-clicking on them shows their parameters. The "swallow" block is set to collect the last 10,000 values of the swallow output from the $\Sigma\Delta$ modulator. At $F_{ref} = 10^7$ Hz, 1 ms of simulation time is required for a completely new data set. When the simulation is paused, an array named swallow is accessible from the MATLAB command window. Various MATLAB commands can be used to analyze the data:

- "size(swallow)" will return " 10,000   1 " after the first millisecond.
- "mean(swallow)", "var(swallow)", and "var(abs(swallow))" will return the average value, the variance, and the variance of the absolute value, respectively.
- "max(swallow)" and "min(swallow)" will give the extreme values of swallow.
- "ssw = sort(swallow);" will sort swallow in ascending order—do not forget the semicolon unless you want all 10,000 values to be displayed on your screen—and name it ssw. Then you can use "ssw(1:10)" to see the first 10 values, for example.

The same processes can be used on dphase, which shows the digitally integrated value of swallow, which indicates the phase deviation. These derived characteristics are used in multiple places, for example, in Fig. U.2, and in comparing the range of carry values for different architectures.

6.7.4 Log Plots

The "TVwide2" block in the lower right can be used to acquire data to create PPSD plots with log-frequency axes using the MATLAB program's plot capabilities. Similar data from different models can also be compared. "TVwide2" is a modified version of the "TVwide" SA, where the "Frame Scope" within the "Spectrum Scope" has been replaced by a block that sends data to the workspace (like the swallow block). The data are stored under the name tunev in an $a \times 1 \times b$ matrix.

The contents of the buffer at the end of the bth frame are stored in a:1:b, which can be acquired when the simulation is stopped or paused. In this case, the buffer of size 4096/crs must be filled at the sampling rate of 2.56 BW, producing four frames with a 0.75 overlap. Therefore, each frame takes 400 cycle/(BW×crs) in simulation time (e.g., 400 μs for a 1 MHz BW at crs = 1). Double-click the block to set the bandwidth BW. The time to fill the buffer and the noise bandwidth are computed from BW and displayed on the block.

The MATLAB script specplot.m (type specplot in the MATLAB command window) will take the FPSD data from the tuning line, divide it by f_m^2 to convert it to PPSD, and create the graph. The last frame created will be plotted, but the user can then choose an earlier frame. The plotted data are named ypsd. You can save ypsd under another name (e.g., ypsd1 = ypsd) for later use. One reason to do that is to plot it on the same graph with data from another simulation that has the same x-axis. The MATLAB script specomp.m will make multiple log plots (type specomp in the command window). The user provides the names of the saved files (e.g., ypsd1) to plot. (By default, it will plot the data named ypsd if not directed otherwise.)

If we right-click on "TV wide2" and choose to "Look Under Mask," we see the block diagram of a typical spectrum analyzer except that the Spectrum Scope has an unusual look. Its mask has been edited (right-click, Edit Mask) to show the word "data" and an arrow. If we look under this mask, we see the diagram of a Spectrum Scope except that the normal "Frame Scope" has been replaced by a data acquisition block such as is described in the previous section. This is a makeshift modification to a spectrum analyzer to allow the data to be acquired rather than displayed and thus to permit an eventual log-frequency plot. It uses the usual components of the SA to perform the DFT on the data. If we double-click on the modified Spectrum Scope, we see the usual four-tab window. Some of the parameters in the window apply (e.g., "Buffer size"), but the parameters in the last three tabs no longer apply to anything. Perhaps it is not surprising that this is one of the blocks that did not function properly when the model was run on an updated version of the Simulink program. However, replacing some of the internal blocks with their updated versions from the Simulink Library solved the problem.

6.8 CPandITrunc.mdl

This model is the same as CPandIplus.mdl except that the MASH modulator has been modified to incorporate truncation circuits between the first and second modulators and between the second and third modulators, as described in Section 5.8.2. Double-click the "Truncated MASH modulator" and then Truncate2 or Truncate3 to see how Fig. 5.25 is realized in the model. The values $n(2)$ and $n(3)$ in $2^{n(m)}$ in Fig. 5.25 are set by variables trunc2 and trunc3, respectively, in the Model Workspace (Control-H). To prevent truncation, set the variable equal to the number of accumulator bits or greater.

6.9 ADAPTING A MODEL

Let us consider the procedure we may follow to adapt `CPandITrunc.mdl` to experiments with dither. Open `CPandITrunc.mdl` and `Dither.mdl`. Save `Dither.mdl` as `DitherTrunc.mdl` and then delete its ΣΔ modulator ("Digital Phase Compensator"). Copy the ΣΔ modulator ("Truncated MASH modulator") from `CandITrunc.mdl`, paste it in place in `DitherTrunc.mdl`, and reconnect it to the rest of the model, deleting unneeded inputs. Open Model Explorer (`View/ Model Explorer` or `Control-H`) and from within Model Explorer, open the `Model Workspace` for `CPandITrunc.mdl` (from the file list on the left of the window). Copy the `Trunc2` and `Trunc3` MATLAB variables, open the `Model Workspace` for `DitherTrunc.mdl` and paste them in that workspace. Then run `DitherTrunc.mdl` without dither to verify that you see the same results that are obtained when `CPandITrunc.mdl` is run.

Next we briefly consider how the model for the SP-MASH synthesizer, `SPmash. mdl`, was created, when the paper describing that configuration appeared, as this book was being finalized. Beginning with the `CPandI.mdl`, its modulator was modified to reflect Fig. 3.17. The changes can be seen by comparing the two models. The length of the first accumulator is again determined by the LSB in n_{frac} and the length of the other accumulators is set to $1/N_a = 2^{-bits}$, as determined by the variable `bits` in the Model Workspace. The seed is expected to be 0 in this design, so it was set to 0 in the workspace and the display in the model that had formerly shown `seed` was changed to display `bits`.

6.10 EFeedback.mdl

In `EFeedback.mdl`, the ΣΔ modulator is changed to a third-order error feedback modulator (EFM), as described in Section 5.2 (Fig. 5.1). The quantizer for this model is shown in Fig. 6.10. The Simulink Quantizer block rounds to the nearest number, so it functions as a midtread quantizer, as shown in Fig. 6.11. However, this is not a proper simulation of a hardware quantizer for negative inputs. For example, the output with -0.5 input should be -1, not 0, as shown in Fig. 6.11. The tread at 0 is wider than the other treads by 1 LSB. If the input word is 3 bits wide, for example, the tread at 0 input is 1.125 wide, whereas the other treads are 1 wide. The difference becomes

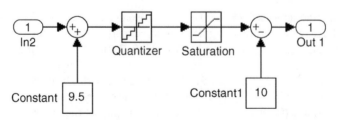

FIGURE 6.10 Quantizer model.

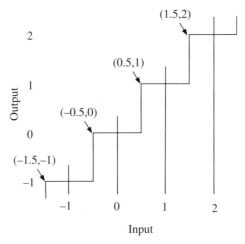

FIGURE 6.11 Simulink Quantizer block.

smaller with longer words, but it is still not correct. To make the simulation accurate, we operate with positive inputs to the quantizer block by offsetting the input up and then offsetting the output down. This can be seen in Fig. 6.10. There, however, the offset at the input is lower than the offset at the output by 0.5. This changes the quantizer to a midriser quantizer, as illustrated in Fig. 6.12.

The script `FBall.m` [`FBall(start_fraction, number_of_frac-tions, end_fraction)`] produced Figs. 5.3–5.6, whereas `FBone.m`

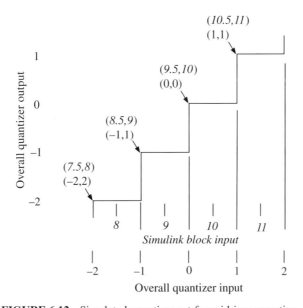

FIGURE 6.12 Simulated quantizer set for midriser operation.

[FBone(fraction)] produces similar information at a single fractional input. The number of bits, the quantizer limits, and the offset (the input offset minus the output offset) can be changed in the first part of the script when desired. A model of the modulator can also be tested; FBmodTest.mdl permits observation of the quantizer input and output on a scope as well as acquisition of these variables as data. Because the script uses the MATLAB function floor rather than round, the problem experienced by the Simulink Quantizer block does not occur.

6.11 FeedForward.mdl

In FeedForward.mdl, the $\Sigma\Delta$ modulator is changed to a third-order feedforward modulator (RS&A), as discussed in Section 5.3 (Fig. 5.7). The quantizer for this model is the same as we just discussed (Figs. 6.10 and 6.12).

The script FFall.m [FFall(start_fraction, number_of_fractions, end_fraction)] produced Figs. 5.10–5.13, whereas FFone.m [FFone(fraction)] produces similar information at a single fractional input. Again, the number of bits, the quantizer limits, and the offset (the input offset minus the output offset) can be changed in the first part of the script. A model of the modulator can also be tested; FFmodTest.mdl permits observation of the quantizer input and output on a scope as well as acquisition of these variables as data. A spectrum analyzer, which produced Fig. 5.14 (each of the PSDs separately), has been added to this model.

6.12 MASH MODULATOR SCRIPTS

The scripts mashone.m, mashall.m, mashall2.m, mashall3.m and mashall4.m simulate MASH modulators through fourth order. Details are listed in Section M.8.

All these scripts output the minimum, average (mean), and maximum modulator outputs and the variance of the output for each value of n_{fract} simulated. Other outputs are provided by the various scripts. The command mashall(order, fstart, fpts, fstop, bits, m) simulates a MASH modulator of the specified order at a number of evenly spaced values of n_{fract} equal to fpts, starting at fstart (typically 0) and ending at fstop (typically 1). Each modulator has a number of bits given by bits. The number of clock cycles simulated at each n_{fract} is 2^m. The other scripts use bits $= 14$ and m $= 15$ by default, so these values must be altered within these scripts if required. Specifying parameters within the command increases versatility, but can be inefficient when the command must be typed multiple times with the same parameters. Similarly, unneeded outputs can decrease efficiency. These are some of the reasons for multiple scripts.

The command mashone(order, n_{fract}) simulates a MASH modulator at the one specified n_{fract} input. All the scripts initially set the least significant bit in the first accumulator.

6.13 SynStep__.mdl

This set of models is used for step response simulations in Appendix H. `SynStepSH.mdl` uses a mathematically modeled PD and a switch driving a mathematical integrator, followed by an ideal S&H circuit that can be bypassed. With the S&H bypassed, it produces the response of Section H.1.1 and Fig. H.1. With the S&H in place, it produces the response of Section H.1.2 and Fig. H.2. `SynStepSH2.mdl` can produce the same responses but it has a PD built from blocks that more closely relate to hardware, which are required for the next two versions.

`SynStepSH3.mdl` has a S&H such as shown in Figs. 2.5 and 2.6. It produced the plots in Figs. H.3–H.6, which are discussed in Section H.1.3.

`SynStepSH4.mdl` contains the improved S&H discussed in Section H.1.4 and it produced Figs. H.9 and H.10.

The displays produced by these models are obtained by double-clicking on the scopes.

6.14 OTHER METHODS

Although lacking the versatility of Simulink models, simulation programs such as are discussed in Chapter F.12, including MATLAB scripts, can produce some types of information, including transient responses, much faster than can a Simulink model. Executing `SynCP1012.m` (`SynCP.m` with parameters for Fig. F.10.12) to produce Fig. F.10.12 requires less than 2 s on a 2.1 GHz computer.[43] Producing the plot in part b of Fig. F.10.12 with `SynStepSH.mdl` requires 32 s on the same computer. Of course, either method can be slowed by the way it is executed; `SynStepSH.mdl` took 45 min when a scope sampling rate was set too high.

We have also shown the use of MATLAB scripts for simulating the purely digital $\Sigma\Delta$ modulators. Scripts for the simulation of many other $\Sigma\Delta$ modulators, with multiple data presentations, are available on the CD accompanying Crawford [2008] and these can sometimes be easily modified to represent other modulators. As an example, Fig. 6.13 shows the PPSDs (before multiplication by NH) resulting from the RS&A feedforward modulator whose spectrum is shown in Fig. 5.14, but on a log-frequency plot (and for a simulation with more bits). It was obtained by a slight modification of Crawford's script (U12718_order3_ ffdsm.m) for a type-3 Jackson feedforward modulator. The same PPSD, but plotted linearly against frequency, could be obtained from the lowest spectrum analyzer ("PPSD from sequence") in the FeedForward.mdl.

Bizjak et al. [2008] claim rapid simulation ["... 10 seconds, instead of the approximately 10 hours that would be typically required for transistor-level simulations of the whole PLL ..."] using Simulink models, but processing output data subsequent to the simulation using the MATLAB program. The MATLAB processing time was not considered. They converted tuning voltage to phase without going through the intermediary step of creating a sinusoidal output signal, an approach that is also employed in `SynCP.m` (Chapter F.12). This would seem

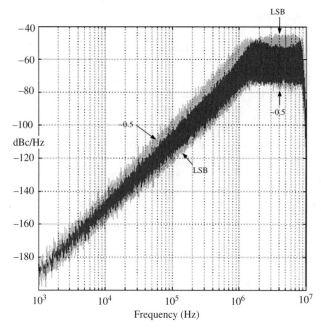

FIGURE 6.13 Superimposed PPSDs with two inputs to RS&A feedforward modulator, from a modification of a script by Crawford.

to provide a significant computational simplification at the expense of a loss of a true output spectrum in regions where $\mathcal{L}$ differs from L. Plans for such simulations are discussed in Section A.2.4.

De Muer and Steyaert [2003, Section III] claim much faster analysis ("... a full nonlinear analysis in approximately 15 seconds...") by analyzing the nonlinear loop circuitry in open loop, characterizing the noise spectrum of the CP pulses, and applying the closed loop response of a linear loop to that noise. This facilitates detailed investigation of the nonlinear effects at varying operating points. Part of the speed increase is attributed to the use of a discrete time analysis with the assumption that "All sampling, except in the prescaler is assumed to be at the edges of the reference frequency signal" (the prescaler operates at a higher frequency). Unfortunately, this would seem to mask the effects of varying sampling rate (Section 2.3.3).

CHAPTER 7

DIOPHANTINE SYNTHESIZER

Another method that has long been used to achieve fine frequency resolution simultaneously with wide loop bandwidth is the combining of two synthesizer PLLs. One method for doing this, which we discussed in Section F.8.2.3, is an architecture that we called Offset Reference after Smith [1996]. It has an advantage over other dual loops (Section F.8.2.2) in which the output frequency of one loop is divided before that output is mixed with the other, because the output frequencies of these individual loops tend to be more commensurate and, therefore, more easily mixed. (When one frequency is much smaller than the other, it is difficult to separate the mixing products from each other.) We found that developing a tuning algorithm for the Offset Reference loop can require some effort. A synthesis method [Sotiriadis, 2006] called diophantine, after the mathematics that has been applied to it, includes Offset Reference as a subset. It provides a mathematical basis that expands the design possibilities and facilitates the specification of design parameters and of a tuning algorithm. In addition, it has been applied to synthesizer architectures for more than two loops [Sotiriadis, 2008a].

Figure 7.1 shows one loop of a diophantine synthesizer. Other loops would be fed by the same basic reference frequency $F_{\text{ref},0}$ and their output frequencies would all be added or subtracted in frequency mixers.

Advanced Frequency Synthesis by Phase Lock, First Edition. William F. Egan.
© 2011 John Wiley & Sons, Inc. Published 2011 by John Wiley & Sons, Inc.

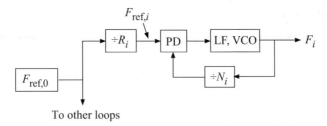

FIGURE 7.1   One synthesizer loop in a diophantine synthesizer.

7.1 TWO-LOOP SYNTHESIZER

Figure 7.2 shows a two-loop diophantine synthesizer while revealing some of the mathematical relationships between parameters. Sotiriadis has shown that divide ratios N_1 and N_2 can be found that generate any frequency multiple of Δf in Fig. 7.2. This should come as no surprise after the description and example in Section F.8.2.3, but he has expanded the available design choices in a way that can be significant, as we shall see. The frequencies F_1 and F_2 are the synthesized frequencies that are added. In this configuration, their sum F_{out} is the frequency of one of the VCOs. This method of adding the frequencies has some advantages, which we will discuss. The synthesis method generates the frequencies F_i, which are added, either directly or, as shown in Fig. 7.2, in a loop.

From Fig. 7.2 we can write

$$F_{out} = F_1 + F_2 \tag{7.1}$$

$$= (R_2 N_1 + R_1 N_2)\Delta f. \tag{7.2}$$

Looking at the example in Chapter F.8, in Fig. F.8.12, $R_1 = 255$ and $R_2 = 256$, so we can write Eq. (7.2) for that design as

$$F_{out} = (256 N_1 + 255 N_2)\Delta f \tag{7.3}$$

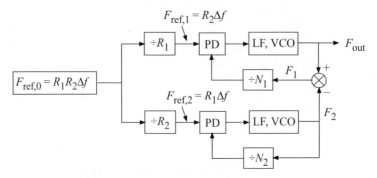

FIGURE 7.2   Two-loop diophantine synthesizer.

and we can write a frequency step as

$$\Delta f_{\text{out}} = (256\Delta N_1 + 255\Delta N_2)\Delta f \tag{7.4}$$

so that if

$$\Delta N_2 = -\Delta N_1, \tag{7.5}$$

the change in the output frequency will be

$$\Delta f_{\text{out}} = (\Delta N_1)\Delta f, \tag{7.6}$$

giving the synthesizer a step size of Δf, even though the two reference frequencies $F_{\text{ref},i}$ are much higher than Δf, thus allowing the loops to have much greater bandwidths. Sotiriadis has also found that the output frequency range

$$\text{Range}(F_{\text{out}}) = \overline{F}_{\text{out}} \pm F_{\text{ref},0} \tag{7.7}$$

can be generated by mixing two synthesized frequencies F_1 and F_2, each with a range

$$\text{Range}(F_i) = \overline{F}_i \pm F_{\text{ref},0}, \tag{7.8}$$

where $i = 1$ or 2 (or higher for more than two loops, as discussed below) and the over bars indicate the means of the ranges. We can put this in another way, writing that

$$-F_{\text{ref},0} \le (F_{\text{out}} - \overline{F}_{\text{out}}) \le F_{\text{ref},0} \tag{7.9}$$

can be synthesized using individual loop output frequencies F_i with spans given by

$$-F_{\text{ref},0} \le (F_i - \overline{F}_i) \le F_{\text{ref},0}, \quad i = 1, 2, \dots. \tag{7.10}$$

Dividing each variable in Eq. (7.10) by $F_{\text{ref},i}$, this also can be written

$$-R_i \le (N_i - \overline{N}_i) \le R_i. \tag{7.11}$$

Here R_i is the reference divider ratio for loop i. A requirement for these spans to be synthesized is that all pairs of R_i are relatively prime (their greatest common divisor is 1; no number but 1 can divide into both without a remainder). This is true of the values used in Eqs. (7.3) and (7.4). Although there is only one pair under consideration now, the same requirement holds for multiple loops to be discussed below.

Consider again the example in Chapter F.8, Fig. F.8.12 and Tables F.8.3 and F.8.4. Equations (7.7) and (7.8) or Eqs. (7.9) and (7.10) imply a total span for the output and for each of the oscillators of

$$2F_{\text{ref},0} = 2(255)(256)\Delta f = 130,560\Delta f. \tag{7.12}$$

As can be seen from Table F.8.4, the spans there are slightly smaller since the span of N was limited to 500 for each oscillator, whereas Eq. (7.11) calls for

$$\Delta N_i \leq 2(R_i) + 1, \tag{7.13}$$

which would be 511 and 513, respectively, for the two loops. Also, in Table F.8.4, the span of F_{out} is slightly smaller than the spans of either F_i.

An advantage of designing for the maximum span of Eq. (7.9) is that a tuning algorithm exists, as executed in the MATLAB® script loop2tune.m (see Section 7.3). See examples of two-loop diophantine synthesizers in Sotiriadis [2008b, 2008c].

7.2 MULTILOOP SYNTHESIZERS

More than two loops can be combined. The reference frequency is

$$F_{ref,0} = R\Delta f_{out}, \tag{7.14}$$

where

$$R \triangleq R_1 R_2 \cdots R_n \tag{7.15}$$

and R_i is the reference divider for loop i, as it is for $i = 1$ and 2 in Fig. 7.2. If the output frequency span is restricted to that given by Eq. (7.9), the individual oscillator frequencies will have spans given by Eq. (7.10), which is equivalent to Eq. (7.11), and the tuning algorithm is given by the MATLAB script loopxtune.m. Multiple loops can achieve very fine tuning without sacrificing loop bandwidth.

The frequencies can be combined, mixing two signals at a time [Sotiriadis, 2008a], the frequency produced by mixing one pair then being mixed with the next frequency or with a frequency that has been obtained by mixing another pair. The tuning algorithm minimizes the range of the intermediate frequencies. It is important to choose the mean frequencies to avoid spurious mixer responses, and mixing only two signals at a time aids this by allowing filtering after each mixer and by allowing a choice of the order of mixing.

7.3 MATLAB SCRIPTS

7.3.1 loop2tune

If we type loop2tune in MATLAB command window (with that script in the workspace), we will be asked to input each reference divider ratio. For the problem in Chapter F.8, we would input 255 and 256. The response would be

```
R(1)=255, R(2)=256, g=1, z(1)=1, z(2)=-1.
```

$R(1)$ and $R(2)$ are confirmation of our input divider ratios and g is their highest common denominator. The z's are used internally in the algorithm; we should not need them unless we are checking the algorithm. The highest common denominator g should be 1, or we will be told to reenter the divider ratios (e.g., if we had entered 254 and 256, g would be 2; these are not acceptable values). Then we will be told to input a within $\pm 65{,}280$, the output span. That will determine the synthesized frequency,

$$F_{\text{out}} = \overline{F}_{\text{out}} + a\Delta f. \tag{7.16}$$

When we input a, the program will give us $x(1)$ and $x(2)$, where

$$x(i) = N_i - \overline{N}_i. \tag{7.17}$$

It will also repeat our input and a computed value for a to confirm that it is the same as what we did input. The frequency for loop i will be

$$F_i = R_1 R_2 \Delta f \frac{N_i}{R_i} \tag{7.18}$$

$$= R_1 R_2 \Delta f \frac{x(i) + \overline{N}_i}{R_i} \tag{7.19}$$

$$= x(i) R_{j \neq i} \Delta f + \overline{F}_i. \tag{7.20}$$

The frequency F_{out} can be synthesized by summing F_1 and F_2,

$$F_{\text{out}} = [x(1)R_2 + x(2)R_1]\Delta f + \overline{F}_1 + \overline{F}_2, \tag{7.21}$$

or by changing the sign, on $x(2)$ for example,

$$F_2 = -x(2)R_1\Delta f + \overline{F}_2, \tag{7.22}$$

and subtracting F_2 from F_1,

$$F_{\text{out}} = F_1 - F_2 = [x(1)R_2 + x(2)R_1]\Delta f + \overline{F}_1 - \overline{F}_2. \tag{7.23}$$

EXAMPLE 7.1

After we obtain the values of R_i as described above, we want to generate a frequency that is 4000 steps below band center, so we input $a = -4000$ to `loop2tune`.
The program outputs: `a = -4000, x(1) = -175, x(2) = 160`.

From Eq. (7.17), we set the dividers at $N_1 = \overline{N}_1 - 175$, $N_2 = \overline{N}_2 + 160$.

As a result, from Eq. (7.20), $F_1 = -175(256)\Delta f + \overline{F}_1 = 44,800\Delta f + \overline{F}_1$ and $F_2 = 160(255)\Delta f + \overline{F}_2 = 42,330\Delta f + \overline{F}_2$.

From Eq. (7.21) or Fig. 7.2, $F_{\text{out}} = [-175(256) + 160(255)]\, \Delta f + \overline{F}_1 + \overline{F}_2 = -4000\Delta f + \overline{F}_1 + \overline{F}_2$, so we have generated a frequency that is 4000 steps below band center, as requested by our input to `loop2tune`.

If we design for a step size of 10 Hz, for example, $F_1 = 448\,\text{kHz} + \overline{F}_1$, $F_2 = 423.3\,\text{kHz} + \overline{F}_2$, $F_{\text{out}} = \overline{F}_1 + \overline{F}_2 - 40\,\text{kHz}$.

The basic reference frequency in Fig. 7.2 is $F_{\text{ref},0} = (255)(256)10\,\text{Hz} = 652.8\,\text{kHz}$ and the output frequency range is, from Eq. (7.7), $\overline{F}_{\text{out}} \pm 652.8\,\text{kHz}$. This is the range of the output oscillator in the configuration of Fig. 7.2. From Eq. (7.8), the other oscillator, and the IF, have the same spans.

7.3.2 loopxtune

If we type `loopxtune`, we will be asked for a vector of R. If we input [32 43 54] for a three-loop synthesizer, we will be told that two of the R-values are divisible by the same number and to try again. If we then input [32 43 55], we will be given the z-values again and a check that the greatest common denominator is 1 and we will be told to input a within $\pm 75,680$. When we do, we will be given the three values of $x(i)$. Equations (7.16) and (7.17) apply again, while Eqs. (7.18)–(7.21) are modified appropriately to

$$F_i = R\Delta f \frac{N_i}{R_i}, \qquad (7.24)$$

$$= R\Delta f \frac{x(i) + \overline{N}_i}{R_i}, \qquad (7.25)$$

$$= x(i) \frac{R}{R_i} \Delta f + \overline{F}_i, \qquad (7.26)$$

and

$$F_{\text{out}} = R\Delta f \sum_i \frac{x(i)}{R_i} + \sum_i \overline{F}_i. \qquad (7.27)$$

The synthesized frequency is obtained by adding all the frequencies F_i together. As before, if we subtract a frequency, we change the sign on the corresponding $x(i)$.

7.3.3 Algorithm

The algorithm can be seen in the scripts and within the paper by Sotiriadis [2006] (which is also referenced therein). We would build it into the control software in a system to turn the commanded frequency into a set of values N_i.

7.4 SIGNAL MIXING

Let us look at the mixing process for Fig. 7.2, with $R_1 = 255$ and $R_2 = 256$ (which is also Fig. F.8.12) and the frequencies given in Tables F.8.3 and F.8.4, which illustrates some of the consideration for both double and multiloop synthesizers. The frequency spans in that example are nearly the maximums given above. The smallest span is that of F_{out}, and it is within 4% of maximum, so the considerations involved in frequency conversion will differ little from a case where the maximum spans are employed. Of greatest concern is a -2 by 3 (LO multiple by RF multiple) spur. (Table F.8.1 contains information about mixer spurs that was used for preliminary considerations in choosing this architecture.)

Figure 7.3 shows a normalized spur plot (Section F.1.2.1 and Egan [2003]). The synthesized frequency F_{out} is the LO (high-power) input to the mixer, F_2 is the RF (lower power) input to the mixer, and the IF (mixer output) is F_1 in Fig. 7.2 (and in Fig. F.8.12). Table 7.1 summarizes the frequencies (from Tables F.8.3 and F.8.4).

Rectangle 1 represents the frequency ranges of the mixer inputs at the maximum IF, rectangle 2 is at the minimum IF, and rectangle 3 is at the mean IF. These rectangles represent the total ranges of the RF and LO at three IFs, but only frequency pairs that lie on the 1×1 line, and that are specified by the tuning algorithm, actually occur. If the frequency pairs represented by Tables F.8.3 and F.8.4, or those given by loop2tune, were plotted, all the points would lie on the 1×1 line. As the synthesizer progresses through its frequencies in order, the locus on the 1×1 line bounces around because while the LO is progressing uniformly, the IF changes, so the vertical displacement does not advance uniformly. (It would if we looked just at the values for a particular IF.)

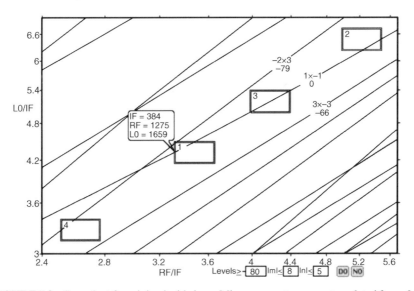

FIGURE 7.3 Spur chart for mixing inside loop. Mixer parameters are extrapolated for an M1 mixer at 12 dBm LO and -16 dBm RF input.

TABLE 7.1 Frequencies of Synthesizer in Fig. F.8.12 in Units of 1000Δ, Where Δ Is the Synthesizer Resolution (Step Size)

	Loop 1	Loop 2	Output
	IF	RF	LO
Minimum	256.0	1275.0	1596.0
Maximum	384.0	1402.5	1721.7
Mean	320.0	1338.8	1658.9
Span	128.0	127.5	125.7

To see what spurious IFs might be produced, we can move from a point on the 1×1 line toward the origin $(0, 0)$. This is the locus of points with the same RF and LO frequencies but with higher IFs. (We could also move in the opposite direction, but there do not appear to be close spurious curves in that direction.) An intersection of that locus with a spur curve indicates that a pair of frequencies that we used to generate the desired IF also can produce another, spurious, IF. The shortest distance to a spur curve along that locus is from the intersection of the 1×-1 line and the left side of rectangle 1—call that point P—to the -2×3 line. There may not be an operating point just at P, but, looking at the plot, it appears to be the worst case, and it is easier to work with that interaction than to search for the nearest operating point. Since the rectangles also move toward the origin as the IF is increased, we can use rectangle 4 to see where a line from P toward the origin would intersect the -2×3 line. At P, the bubble in Fig. 7.3 shows $F_{RF} = 1275$ and $F_{LO} = 1659$ in units of $1000\Delta f$.[44] The -2×3 spur will produce an IF of $(-2F_{LO} + 3F_{RF}) = 507$ at these frequencies. This is the IF of rectangle 4. The 1×-1 line intersects rectangle 3 at P and the -2×3 line intersects rectangle 4 at the same point. The guard band (between the spur and the IF band) width is $(507 - 384) \times 1000\Delta f = 123{,}000\Delta f$ and the required IF filter half bandwidth is $64{,}000\Delta f$ (half of the IF span for loop 1 in Table 7.1), so the required IF filter shape factor is 2.9 $[= (123 + 64)/64]$.

To see how pessimistic is our assumption that an operating point occurs at the left edge of rectangle 1 in Fig. 7.3, we can search all the frequencies produced by `loop2tune`, using the script `loop2spurs`. (Now we are employing the slightly larger bands given by the diophantine algorithm, still centered at the same frequencies; the shape factor obtained above changes slightly to 2.8.) We find that the RF/IF value along the x-axis in Fig. 7.3 (modified slightly to use the full range of the algorithm) never goes below 3.486, so the leftmost operating point is actually where the 1×-1 curve intersects the top of rectangle 1. The shape factor here is a more generous 3.8 $[= 1 + (568.19 - 385.28)/65.28]$. It is possible to argue that this is the minimum shape factor by considering how Figure 7.3 changes when rectangles are drawn for other IFs, but, if this is not apparent, we can verify, again using the script `loop2spurs`, that the lowest frequency of the -2×3 spur, for all synthesized frequencies, is indeed $568{,}190\Delta f$.

What harm can the spur do if it is not adequately filtered? PM sidebands can be induced in the synthesizer output (by a process described in Section F.8.1.2). The spur that is offset by $123,000\Delta f$ will be converted, at the divider input, to phase modulation at that frequency. The divider output phase is sampled at $F_{ref,1} = 256\Delta f$, so aliasing will produce many modulation frequencies within the sampled phase at $f_m = |123,000 \pm n256|\Delta f$, for all values of n (Section F.3.5). These modulations are effectively injected at F_1 and will appear at F_{out} after multiplication by $H(f_m)$. As F_{out} varies, some of these frequencies will fall within the bandwidth of the upper loop and will not be attenuated. In our example, the -79 dB spur will cause -85 dB modulation sidebands (6 dB reduction; see Section F.3.2). These may be small enough for a given system requirement without filtering. If the strength of the weaker mixer input were to increase, the spur amplitude would increase at thrice the rate of that input (in dB); so the sideband levels, relative to the output, would increase 2 dB for each decibel increase in that input and vice versa should the level drop. Also, note that the spur level will vary with different mixers and frequencies.

The spur plot of Fig. 7.3 represents mixing within the loop, as shown in Figure 7.2. The output frequency is still the sum of the two frequencies F_1 and F_2 that are input to the dividers $\div N_1$ and $\div N_2$, respectively, as if the frequencies had each been generated by a VCO and mixed outside the loop (as in Fig. F.8.11). Figure 7.4 shows the spur chart for addition by mixing outside the loops. Table 7.1 applies again except that the first column is now RF, the second is LO, and the third is IF. We see that there is a crossover spur within the range of frequencies used in synthesis. This is the same order and estimated level as in the first case, but it cannot be filtered out. If this level is too large, the level of the RF input would have to be reduced. (In this case, there will not be conversion to PM and corresponding 6 dB reduction.) Since the output is now the IF,

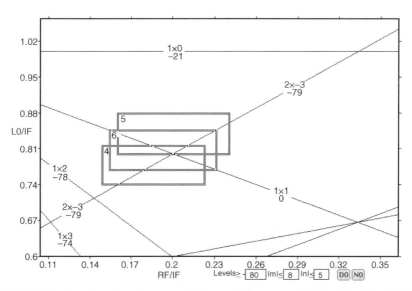

FIGURE 7.4 Spur chart for mixing outside loops. M1 mixer at 12 dBm LO with -16 dBm RF input.

as the frequency progresses uniformly, the rectangles will take very small steps from #4 toward #5 in Fig. 7.4, while the operating point will be somewhere on the 1×1 line within each of those rectangles.

We may try to avoid the crossover spur by changing the median frequencies of the two loops in this second case, although both of the conversions that we have considered are within generally advantageous regions of the spur chart (regions 2 and 3, respectively, in Fig. 7.29 of Egan [2003]).

7.5 REFERENCE-FREQUENCY COUPLING

In the Offset Reference synthesizer that we have been considering, $F_{ref,1}$ and $F_{ref,2}$ differ only by the output step size Δf. Coupling between these frequencies at either PD has the potential to cause phase modulation at a modulation frequency Δf. Since we use this configuration in order to allow the loop bandwidths to be wide compared to Δf, any such modulation would appear at the synthesizer output, unattenuated and amplified by N. Since the diophantine algorithm allow us to accomplish our desired synthesis without so restricting the values of $F_{ref,1}$ and $F_{ref,2}$, we will now consider how we might use it to provide the same output signal while alleviating the potential coupling problem.

To get the same output resolution and range, we need the same values of Δf and $F_{ref,0}$ [see Eq. (7.7)]. Therefore, in our new design (refer again to Fig. 7.2), we again desire

$$R_1 R_2 = 255 \times 256 = 65,280. \qquad (7.28)$$

We want the frequency of PM due to cross-coupling to be as high as possible, but we must also account for aliased frequencies that will occur when that frequency exceeds $F_{ref}/2$, because of the sampled nature of the PD. The result of such considerations leads us to $R_1 = 325$ and $R_2 = 201$.[45] The `loop2tune` program indicates that these are acceptable prescaler ratios with a resulting frequency range of plus and minus

$$R_1 R_2 = 325 \times 201 = 65,325, \qquad (7.29)$$

essentially the same as before. Note that we have been forced to reduce one of the loop reference frequencies slightly, which could result in slightly slower overall synthesizer response.

Now the phase modulation frequency in both loops will be $[(325 - 201)\Delta f =]$ $124\Delta f$. This will be $[124/325 =] 0.382$ times the reference frequency in loop 1. In loop 2, the lowest PM frequency, as a result of sampling at the PD rate of $201\Delta f$, will be $[(201 - 124)\Delta f =] 77\Delta f$. This will be $[77/201 =] 0.383$ times the reference frequency of loop 2. In this case, we have chosen R_1 and R_2 to make the lowest modulation frequency the same percentage of the F_{ref} in each loop. If we assume that bandwidth f_L of each loop is less than 10% of its reference frequency, the PM will be reduced by the loop response at or beyond $3.8 f_L$. Of course, we can increase the attenuation by further reducing f_L, which will also allow more effective low-pass filtering.

Since the frequency ranges for the synthesizer output and for the individual loops are essentially as they were, and since we can choose approximately the same center frequencies,[46] there need be no significant change in the design, except for a decreased sensitivity to undesired PM. If we tune 4000 steps below band center, as in the example above, `loop2tune` now gives $x(1)$ and $x(2)$ as -136 and 200, respectively, rather than -175 and 160 as before. Equal steps with opposite signs in each loop no longer produce an equal output step [Eqs. (7.5) and (7.6)], but these new values of N_i again produce $-4000\Delta f$.

7.6 CENTER FREQUENCIES

The center frequency of the synthesizer can be chosen as any multiple of Δf, but center frequencies of the individual loops are restricted to being multiples of $R_i \Delta f$. We can choose integer values of N_i to give approximately the desired loop center frequencies and then compute the synthesizer's center frequency from Eq. (7.2). If we wish, we can change the resulting synthesizer center frequency to some other value by inserting the required frequency change as a change from center frequency in `loop2tune` and applying the obtained values of $x(i)$ as changes to N_i. The new values of N_i will then produce the desired synthesizer center frequency. We can readjust the center frequency values of N_i without changing the synthesizer frequency by, for example, changing N_1 by R_1 while simultaneously changing N_2 by $-R_2$, leaving F_{out} in Eq. (7.2) unchanged.

CHAPTER 8

OPERATION AT EXTREME BANDWIDTHS

The methods that we have been discussing offer the possibility of obtaining fine step sizes while maintaining wide loop bandwidth at the cost of some complexity. Before deciding to use them, we may try to achieve our design goals with a standard integer-N single-loop synthesizer. This could force us to widen the bandwidth to a point where the sampling process significantly affects our representation of the synthesizer loop. Here, we will extend the discussion of operation at these extreme bandwidths that we began in Chapter F.7.

8.1 DETERMINING THE EFFECTS OF SAMPLING

Control systems using sampled data are traditionally analyzed using z-transforms [Franklin et al., 1990], but we can also employ Laplace transforms, if we include additional terms. The open-loop transfer function then becomes

$$G(f_m) = G_0(f_m) + [G_0(f_m + f_s) + G_0(f_m - f_s)] + [G_0(f_m + 2f_s) + G_0(f_m - 2f_s)] + \cdots, \tag{8.1}$$

where f_s is the sampling frequency and $G_0(f_m)$ is the transfer function in the absence of sampling.

In some ways, the use of Laplace transforms can be more convenient and permit a smoother transition from the usual analysis of continuous (not sampled) loops.[47]

Advanced Frequency Synthesis by Phase Lock, First Edition. William F. Egan.
© 2011 John Wiley & Sons, Inc. Published 2011 by John Wiley & Sons, Inc.

Whether we use z-transforms or Laplace transforms with added terms, the theory is based on the assumption of a fixed sampling rate. To that degree, the improved analysis may still involve an approximation, unless we have taken measures to make the sampling rate constant. Accurate *simulation* can account for the varying sample rate, but exact *analysis* (Section F.9.1) with a varying sample rate is difficult and tends to be restricted to simple loops (see Chapter F.9, especially Table F.9.1).

We can account for (fixed rate) sampling in any PLL by including additional terms in the response at frequencies offset by multiples of F_{ref} (Section F.7.4.4). The program Gsmpl facilitates this and enables us to modify our normal design process to see the effects of sampling. For example, we can use it to modify a Bode plot to see how it changes due to sampling. Figures F.7.20 and F.7.21 show how sampling affects Bode plots for two variations of an example design. Gsmpl was also used to create Figs. F.7.24–F.7.28, which show the variations produced by sampling in a manner that permits the information to be applied to a range of designs.

The sampling that we are discussing here occurs when the phase is read by the PD at the output transition of the frequency divider.[48] When a S&H PD is used, the effect of its zero-order hold is approximated as introducing a delay of $T_{ref}/2$ (Section F.7.2). However, we have also shown some circuits where the control signal is subject to an additional delay in order to obtain effectively a constant sampling rate (Section 2.3.1). When we reversed the inputs to the S&H PD, the output of the zero-order hold changed in synchronism with F_{ref}, at some delay after the divider transition. When we resampled the voltage on an integrator following a PFD, in synchronism with F_{ref}, there was a delay after the divider transition. This delay varies, because the sampling instant varies. Accounting for these delays with a phase shift of $-f_m T_d$, where T_d is the delay from the average time of the preceding divider transition, seems like a reasonable approximation. We are now seeing the phase error that was measured at a varying time, change to its new value at a fixed time, and this seems very similar to sampling the phase error at a fixed rate followed by a fixed delay.

Figure F.7.24[49] shows how the open-loop gain and phase are changed by sampling in a second-order loop with a low-pass loop filter and a zero-order hold (i.e., the type of hold used in a S&H PD, see Appendix P), and Fig. F.7.25 shows this when there is no hold. Changes are shown for four ratios of the low-pass pole frequency to the sampling frequency. Figures F.7.26–F.7.28 show the same information for representative third-order loops. For increased clarity, let us review in more detail the information given there on the second-order loop for one of the pole-frequency ratios, one-tenth of the sampling frequency. We will analyze this case with and without a hold.

8.2 A PARTICULAR CASE

Figure 8.1 shows the Bode plot for the linear version (Fig. 1.1) of a second-order loop with a low-pass filter, where sampling is ignored and no approximation is made for a sampler. This plot has a liner x-axis, which is unusual for Bode plots, so the curves may look strange for that reason. The x-axis shows the modulation frequency relative to the sampling frequency (even though the sampling frequency has no effect on this

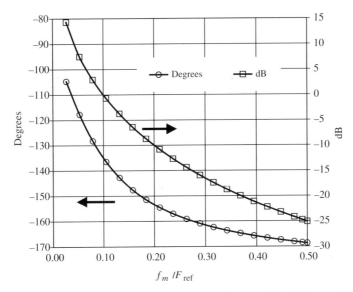

FIGURE 8.1 Bode plot of linear circuit with no sampler, second-order loop with low-pass filter pole at $x = 0.1$.

particular plot). We have adjusted the gain to be unity at the filter corner, where $f_m/F_{ref} = 0.1$, giving a 45° phase margin.[50]

Figure 8.2 shows the *changes due to sampling*, in open-loop gain and phase, for a loop that does not have a hold circuit. At the filter corner, $f_m/F_{ref} = 0.1$, there is only a

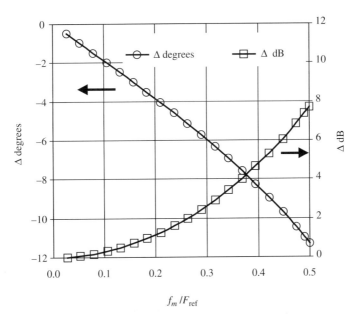

FIGURE 8.2 Changes to the Bode plot in Fig. 8.1 due to sampling, no hold.

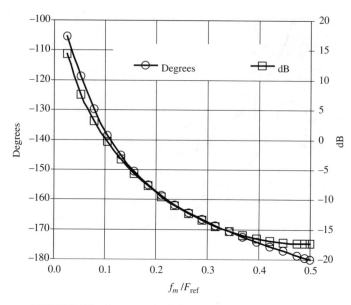

FIGURE 8.3   Response including sampling effects, no hold.

$2°$ reduction in phase margin. The increased gain moves the unity gain frequency higher, causing an additional reduction of only $0.6°$ (based on Fig. 8.1). Therefore, sampling decreases the phase margin from $45°$ to about $42°$, not such a large impact. However, if the loop had less margin to begin with—that is, the unity gain frequency occurred at a higher value of f_m/F_{ref}—we can see that the decrease would be more severe. Theoretically, a second-order loop will not be unstable—it cannot have zero-phase margin at any finite frequency (the phase in Fig. 8.1 does not reach $-180°$ until infinite frequency)—but the sampling effects change that. The *total* response with sampling is shown in Fig. 8.3. The phase just reaches $-180°$ at a finite frequency, half of the sampling rate. It just reaches $-180°$ there but, practically, other phase lags will be added due to stray capacitances and time delays. There is symmetry about $F_{ref}/2$ that requires that the phase shift be $-180°$ there (see Fig. F.7.22).

If the loop has a hold circuit, for example, because a S&H PD is used or because of resampling[51] to avoid added phase noise (Appendix E), the first-order approximation of the effect of the hold circuit at the fundamental frequency (Section F.7.2) is a linear phase shift reaching $-180°$ at F_{ref} and a gain multiplier of $sinc(f_m/F_{ref})$. The changes to the linear loop produced by this approximation are shown in Fig. 8.4. They produce an additional phase lag of $18°$ at $f_m/F_{ref} = 0.1$. The gain reduction of about 0.15 dB˙ there reduces the zero-gain frequency slightly, so the added lag will be closer to $17°$. But that reduces the phase margin to $38°$, which is not very comfortable and increases the overshoots in frequency and step responses. Figure 8.5 shows *additional* changes when the accurate response, obtained by including the added terms at frequencies offset by multiples of F_{ref}, replaces the first-order approximation. They increase the phase and decrease the gain, improving the phase margin relative to this first-order

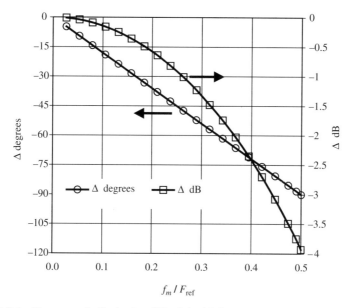

FIGURE 8.4   Changes to the Bode plot of Fig. 8.1 with first-order correction for a hold circuit.

approximation. However, the effect at $f_m/F_{ref} = 0.1$ is very small, which shows the value of the first-order approximation. The effects of sampling at $f_m/F_{ref} = 0.1$ is considerably greater with the hold circuit than it was without, but inclusion of the first-order approximation causes accuracy with a hold to be better than without the hold.

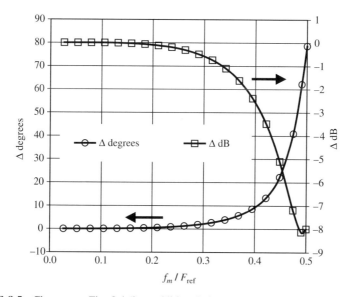

FIGURE 8.5   Changes to Fig. 8.4 (i.e., additional changes to Fig. 8.1) when the effects of sampling are more accurately portrayed.

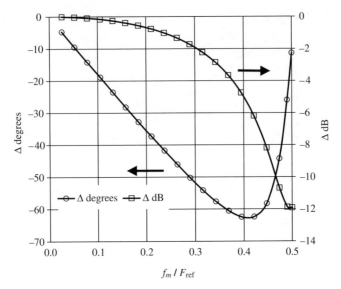

FIGURE 8.6   Changes to Fig. 8.1 due to sampling with a hold present.

Figure 8.6 shows the total change to the simple linear model from the inclusion of the additional terms, and Fig. 8.7 shows the resulting Bode plot with sampling effects included. Note that $-180°$ again occurs at $f_m/F_{ref} = 0.5$, satisfying symmetry requirements, but, this time the phase goes through $-180°$ first at a much lower frequency, giving us a gain margin of only 8 dB.

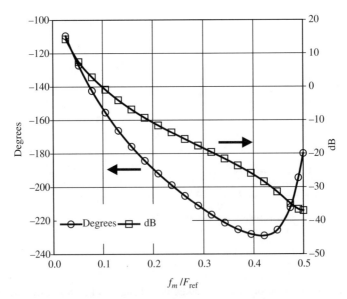

FIGURE 8.7   Bode plot of loop with hold circuit including sampling effects.

Although Figs. F.7.24–F.7.27 show the effect of sampling on gain and phase for certain second- and third-order loops, with and without samplers, as f_L approaches F_{ref} and while these are quintessential PLLs, they do not encompass all possible design variations that we might want to analyze. Gsmpl however can be used to include the effects of sampling for any loop and may be more convenient than using Figs. F.7.24– F.7.27. Figure 8.8 is an updated version of Fig. F.7.24,[49] which includes the information in Figs. 8.4 and 8.5. The update is the same as the version in the current errata for *FS2*, which is necessary due to a sign error in Fig. F.7.24 and which perhaps shows another advantage of using Gsmpl or of at least having it available to check results.

8.3 WHEN ARE SAMPLING EFFECTS IMPORTANT?

We saw, in the examples above, that the effects of sampling on $G(f_m)$ were small at $f_m/F_{\mathrm{ref}} = 0.1$, if we used the first-order approximation for the hold when a hold is present. We could also see that the gain and phase changes due to sampling increase as f_m/F_{ref} increases. Gardner [1980] has studied second- and third-order type-2 PLLs with CP PDs and concluded that troubles begin to appear when sampling effects are ignored and the loop bandwidth is less than approximately 10 times F_{ref}. We have considered first-order type-1 and second-order type-2 loops with and without hold circuits with results shown in Figs. F.7.24–F.7.28, and these results support the same conclusion, with the restriction that the first-order approximation for the hold be employed when the hold is present. However, we can also see that the changes are smallest in some configurations [especially for the type-1 second-order loop without a hold (Fig. F.7.25)] when the low-pass filter is sufficiently low in frequency, say $f_p \leq F_{\mathrm{ref}}/10$. Small changes in gain and phase due to sampling occur even when these limits are met; the fact that they are small does not preclude them from being occasionally important.

8.4 COMPUTER PROGRAM

The current version of Gsmpl, which is available online, now computes $H(f_m)$ and $[1 - H(f_m)]$, as well as $G(f_m)$, with sampling. Appendix G describes the use of Gsmpl and illustrates the creation of Bode and Nyquist plots from the results and shows agreement with stability limits obtained by Gardner [Gardner, 1980, pp. 1849–1958; Gardner, 2005, p. 276).

8.5 SAMPLING EFFECTS IN ΣΔ SYNTHESIZERS

Sampling effects are not likely to be significant in ΣΔ synthesizers because the loop has to be narrow enough, relative to F_{ref}, to enable the quantization noise to be adequately filtered. Lacaita et al. [2007, p. 69] determined that the ratio between F_{ref}

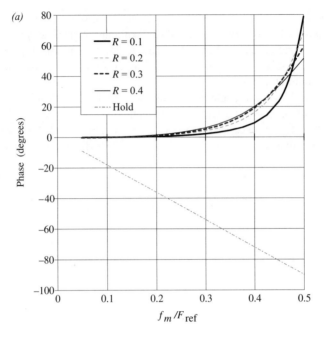

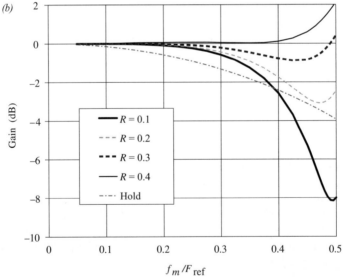

FIGURE 8.8   Revised Fig. F.7.24, corrections to open-loop (*a*) phase and (*b*) gain for four values of $R = f_p/F_{ref}$. For $R = 0.1$, the data are the same as shown in Figs. 8.4 and 8.5.

and the -3 dB bandwidth of five fractional-N synthesizers reported in the literature varied between 200 and 1000. We have given examples (Section 4.1.6, Table 4.2) of four loops where this ratio exceeds 340. These far exceed a ratio of 10, below which we may expect significant effects from sampling. Nevertheless, the loop response, with sampling included, is given in Section Q.4 and evidence of the (normally unimportant) effect of sampling on the out-of-band quantization noise is given in Section Q.5.

CHAPTER 9

ALL-DIGITAL FREQUENCY SYNTHESIZERS

Driven by the advantages of incorporating frequency synthesizers in ICs that are designed for digital circuits, especially those using processes that are not very compatible with analog circuitry, researchers have developed two new architectures that we will discuss here. The Flying Adder is used to generate a desired average frequency, which is not spectrally pure, while the all-digital phase-locked loop (ADPLL) synthesizer attempts to generate a spectrally pure signal.

Great importance is placed on minimizing the use of analog circuitry because it is not easily integrated into modern ICs that are designed for very high digital circuit density (e.g., deep submicron CMOS). Not only do the analog components tend to use relatively large areas but also the voltages used in these circuits are inappropriately low for analog functions, bringing them close to junction voltages and noise levels. Unfortunately, the availability of only discrete timing instances for signal transitions in clocked digital circuits presents a problem for frequency synthesizers.

A common compromise is the synthesis of signals with the desired average frequency, even though transitions deviate from the desired instants by some limited amount. We see this in the fractional-N divider, which presents an average frequency to the phase detector, but there the filtering action of the PLL, aided by cancellation from a DAC or by $\Sigma\Delta$ modulation, reduces the effect, at f_{out}, of the jitter that appears at the PD. By itself, the first-order $\Sigma\Delta$ modulator synthesizes an average frequency, $\overline{f}_{out} = n_{fract}F_{clock}$ when $n_{fract} \leq 0.5$.

We will also see this in the Flying Adder synthesizer. It can generate frequencies that are exact submultiples of a fixed harmonic of the input (reference) frequency.

Advanced Frequency Synthesis by Phase Lock, First Edition. William F. Egan.

And, by using fractional-N techniques, it can generate intermediate frequencies, with the jitter that accompanies these techniques, but it has no loop to reduce that jitter. Nevertheless, there are applications where the accurate average frequency that it produces is sufficient and where, therefore, the application can benefit from its digital nature [Xiu, 2008].

In traditional ADPLLs, such as those discussed in *PLB*, all the signals are synchronized with a clock, except the input signal. The output is a binary-level signal with the same average frequency as the input, but with transitions that must occur in synchronization with a clock rather than with transitions of the input signal. The ADPLL synthesizer that we will consider differs from these in that its output frequency is a multiple, an exact multiple, not just on average, of the reference frequency, even a multiple containing a fraction.

9.1 THE FLYING ADDER SYNTHESIZER

9.1.1 The Concept

Consider the modification of Fig. 5.21 shown in Fig. 9.1a[Xiu, 2008]. The two outputs from the VCO/phase-select mux combination have been switched. The synthesizer's divider is driven by the VCO directly, while the output is selected in the phase-select mux. Thus, a variable frequency can be obtained from a simple synthesizer and the output frequency can be changed rapidly without being restricted by the synthesizer speed. Now something like the glitch in Fig. 5.22 becomes the intended waveform because the objective is no longer to delay the effective edge of the output waveform by an *additional* T_{out}/M but to obtain an edge that is separated from the previous output edge by just $k(T_{out}/M)$, as shown in Fig. 9.2, where $M = 8$ and $k = 2$—now we want the waveforms to respond quickly. Here, the phases are numbered such that Φ_i is T_{out}/M later than Φ_{i-1} and the figure illustrates sequential advances of i by $k = 2$. When the level of Φ_1 rises, the output (which is connected to Φ_1) also rises. This clocks the accumulator and it adds $k = 2$ to its contents. This in turn steps the phase select in the mux by 2, from 1 to 3, causing the output to switch to Φ_3. As a result, the output drops to the level of Φ_3, which is low at that time. Then, 1/4 cycle after the initial positive edge, Φ_3 rises, producing a clock edge separated by 1/4

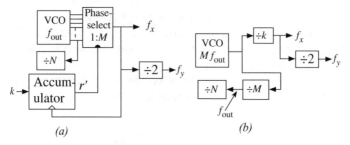

FIGURE 9.1 Flying Adder circuit in (*a*) and equivalent in (*b*).

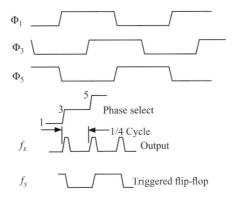

FIGURE 9.2   Flying Adder waveforms: $k = 2$, $M = 8$.

cycle of f_{out}, which is $2(T_{out}/M)$, from the previous output rising edge. The process is then repeated when Φ_5 is connected to the output, producing another cycle with period $2(T_{out}/M)$. These edges trigger a flip-flop, which then has a period of $4(T_{out}/M)$ and a 50% duty cycle.

The circuit of Fig. 9.1b is equivalent to that at Fig. 9.1a. The output at f_x has a period of

$$T_x = kT_{out}/M, \quad 1 \leq k < M, \tag{9.1}$$

but, as before (Section 5.7.1), the actual circuit can have unintended phase modulation due to unequal delays among the phases. We will now consider the output spectrum in the absence of such perturbations.

9.1.2 Frequencies Generated

The number of bits w in k and in the accumulator register can be more than m, the number required to control the mux, where $M = 2^m$. In that case, the m most significant bits comprise r' and k can be written

$$k = W2^{m-w} = K_{int} + k_{fract}, \tag{9.2}$$

where W is a whole number having w bits, K_{int} is a whole number represented by m bits, and $k_{fract} < 1$. For example, if $w = 8$, $m = 3$, and $k = 5.125 = 101.001_2$, $W = 5.125 \times 2^5 = 10,100,100_2$, $K_{int} = 5 = 101_2$, and $k_{fract} = 0.125 = 0.001_2 = 0.00100_2$.

This is similar to a first-order $\Sigma\Delta$ modulator except that the output there is a single bit. It is more similar to the accumulator used in the Digiphase Synthesizer (Section F.8.3.1), where the fractional bits overflowed into the bits that are compared to the $\div N$. The output period is still given by Eq. (9.1), but this becomes the *average* period,

$$\overline{T}_x = \frac{k}{M} T_{out} \quad \text{for} \quad 1 \leq k < M. \tag{9.3}$$

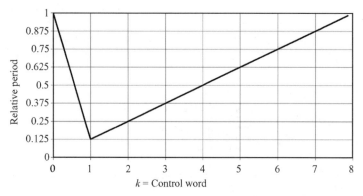

FIGURE 9.3   Ratio $\overline{T}_x/T_{\text{out}}$ versus k, $M = 2^3$, 6 bit register.

The individual periods are either $K_{\text{int}}T_{\text{out}}/M$ or $(K_{\text{int}} + 1)T_{\text{out}}/M$, just as in the fractional-N divider.

A plot of $\overline{T}_x$, the average period, as a function k is shown in Fig. 9.3. We cannot generate a period T_x that is shorter than T_{out}/M. If $k < 1$, r' will either advance by 1, when the fractional part rolls over, or remain unchanged from the previous period. In the latter case, T_x will equal T_{out}, resulting in

$$\frac{\overline{T}_x}{T_{\text{out}}} = \frac{k + (1-k)M}{M} = 1 - k\left(1 - \frac{1}{M}\right) \quad \text{for} \quad 0 \le k \le 1. \tag{9.4}$$

For the example in the figure, this is $1 - (7/8)k$.

The corresponding frequency $\overline{f}_x = 1/\overline{T}_x$ is plotted in Fig. 9.4, where we see that the attainable frequency range is

$$f_{\text{out}} \le \overline{f}_x \le Mf_{\text{out}}. \tag{9.5}$$

The output that is generated when $0 < k < 1$ deviates significantly from a square wave [Sotiriadis, 2010a], being a mix of long and short half periods. Since more uniform outputs can be generated at the same frequencies with $k > 1$, let us restrict our

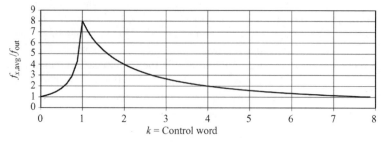

FIGURE 9.4   Ratio of $\overline{f}_x$ to f_{out} versus k, $M = 2^3$, 6 bit register.

commands to $k = 0$ (for which $\bar{f}_x = f_{out}$) and $k \geq 1$. Then, the frequencies are

$$\frac{\bar{f}_x}{f_{out}} = \frac{T_{out}}{\bar{T}_x} = \frac{M}{k} \quad \text{for} \quad 1 \leq k < M. \tag{9.6}$$

Since $\bar{T}_y = 2\bar{T}_x$, the relative period in Fig. 9.3 is also equal to $\bar{T}_y/(2T_{out})$ and the ratio plotted in Fig. 9.4 is also $2\bar{f}_y/f_{out}$. In this region, the spectrum of f_y shows significant power [Sotiriadis, 2010b] at the average frequency,

$$\bar{f}_y = \frac{\bar{f}_x}{2} = f_{out}\frac{M}{2k}, \tag{9.7}$$

plus spurious components at multiples of the reciprocal of the fundamental period [Sotiriadis, 2010a],

$$f_v = f_{out}\frac{M}{W}2^{LFZ}, \quad k_{fract} \neq 0, \tag{9.8}$$

where LFZ is the number of LSBs in k_{fract} that are 0.[52] If $k_{fract} = 0$, the output at f_y is a square wave and the spacing between spurs is half of the value given by Eq. (9.8), so we could also write

$$f_v = f_{out}\frac{M}{W}\frac{2^{LFZ}}{1 + \delta_{0,k_{fract}}}. \tag{9.9}$$

When W is odd, 2^{LFZ} equals 1. As with the fractional-N synthesizer in Section 2.1, the spacing between spurs increases as 2^{LFZ}. (A similar expression could be written for the fractional-N synthesizer.) The average frequency $\bar{f}_y$ is a multiple of f_v.[53]

9.1.3 Jitter

In determining whether the Flying Adder's output signal will satisfy a requirement, it will often be important to know the bounds on the jitter of the signal. Fortunately, Sotiriadis [2010b] has determined these limits for the signal at f_y. If $k_{fract} = 0$, there is no jitter; each cycle is the same as the preceding cycle. Otherwise (for $k > 1$), the greatest difference between the periods of adjacent cycles is

$$\Delta \triangleq \frac{1}{Mf_{out}}, \tag{9.10}$$

the time shift between adjacent phases from the VCO. He also gives values for four other jitter definitions. Perhaps the value of the shortest output period is most important in many applications. It is easy to see that the shortest output period is

$$T_{y,min} = 2K_{int}\Delta, \tag{9.11}$$

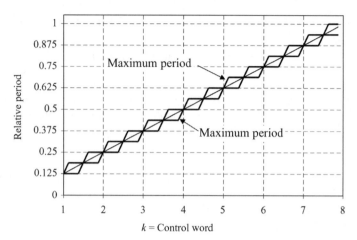

FIGURE 9.5   Maximum, average, and minimum periods relative to $2T_{\text{out}}$ versus k, $M = 2^3$, 6 bit register.

if $k_{\text{fract}} < 0.5$, and if $k_{\text{fract}} \geq 0.5$, it is

$$T_{y,\min} = 2(K_{\text{int}} + 0.5)\Delta. \tag{9.12}$$

These periods are plotted in Fig. 9.5 relative to $2T_{\text{out}}$. It is apparent that the relative variations in period decrease for larger values of k.

   If we are concerned about variations in phase delays, we could decrease Δ in Eqs. (9.11) and (9.12) to account for the expected variation, which we have been ignoring. The time required for M fundamental periods is

$$M/f_v = T_{\text{out}}W2^{-\text{LFZ}}(1 + \delta_{0,k_{\text{fract}}}), \tag{9.13}$$

which is an integer multiple of T_{out}. Therefore, the average frequency over any multiple of this period has the same accuracy as does F_{out}, even if the phase delays are not all equal.

9.1.4 Suppression of Spurs

Xiu [2008] has experimented with suppressing the spurious outputs from the Flying Adder circuit and found success by dithering k_{fract} with random numbers, adding random numbers between $-k_{\text{fract}}$ and $+k_{\text{fract}}$ to k_{fract}. He also used triangular and sawtooth numerical modulation. These tended to produce a low noise plateau, but may be good for spreading the spectrum where that is a desirable feature. He also tried $\Sigma\Delta$ modulation, third and fifth order, without much success. This may suggest that random numbers might be better than $\Sigma\Delta$ modulation in Section 5.7.3, although Xiu's experiments were intended to suppress pattern noise, whereas Riley and

Kostamovaara were attempting to suppress noise due to nonideal phase increments (a problem not addressed in Xiu's paper).

9.1.5 Further Development

Flying Adder circuits with added phase control (modulation) and circuits capable of higher speed operation while avoiding potential race problems (glitches) are discussed in Section A.1.

9.2 ADPLL SYNTHESIZER

Great effort is being expended in the development of all-digital frequency synthesizers.

9.2.1 ADPLL Concept

Figure 9.6 shows the concept of a common realization of the ADPLL synthesizer. The output is generated by a digitally controlled oscillator (DCO). This oscillator is digital because its output is a square wave with binary levels and its input is a number that determines the capacitance in its tank circuit, and thus its frequency. A frequency counter counts the output. This counted value is the phase of the output in cycles and it is subtracted from the ideal value to produce the phase error. The ideal value is obtained by increasing the output, from an accumulator, by N each reference period, where, as usual,

$$N = F_{\text{out}}/F_{\text{ref}}. \tag{9.14}$$

The error is sampled and held every T_{ref} and the held value drives the Loop Filter and provides a tuning number to the DCO. [We may note the similarity of this configuration to the Digiphase Synthesizer (Section F.8.3.1), but the phase detection here is obtained by subtracting two numbers rather than by measuring the difference between their times of occurrence.]

Of course, the ADPLL Synthesizer provides great flexibility because parameters and configurations can be altered with relative ease. The circuits also tend to make use of process such as dynamic element matching (DEM), wherein supposedly identical components are interchanged at a rapid rate to reduce the effects of variations in their values. (Controlling noise produced by the resulting modulations is part of the design process.)

Another form of ADPLL synthesizer architecture is discussed in Section A.2.1.

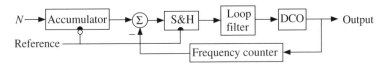

FIGURE 9.6 ADPLL synthesizer.

9.2.2 The Numbers

If the frequency counter has n_{cb} bits (e.g., 10) and the accumulator register has $n_{cb} + n_{fb}$ bits (e.g., 15), the n_{fb} (5) LSBs in the accumulator, and in N, represent a fraction. The value of the MSB in both the accumulator and the counter is $2^{n_{cb}-1}$ (512) cycles and the capacity of the counter is $2^{n_{cb}}$ (1024). The phase error is obtained by modulo $2^{n_{cb}}$ subtraction. If N should equal 200.125, for example, the counter would roll over about every fifth[54] reference instant. If the count in the counter should equal 1023 and the count in the accumulator should be 1.25 at such an instant, the phase error would be represented by $(1.25 - 1023)\mathrm{mod}1024 = (1.25 - 1023 + 1024)$ $= 2.25$. However, the most significant bit of the phase error is taken to be a sign bit, so numbers higher than $2^{n_{cb}-1}$ are interpreted as negative. If the counter output should be 3 rather than 1023, in the above example, the phase error would be $(1.25 - 3)\mathrm{mod}1024 = 1022.25 = 1111111110.01$, which represents -1.75 (a model of this process is shown in Fig. A.6).

9.2.3 Mathematical Representation

The mathematical block diagram (Fig. 9.7) shows the integration of f_{out} to produce φ_{out} and the integration of f_{ref}, followed by multiplication by N, to produce $N\varphi_{ref}$. The phase error φ_e is multiplied by the usual loop parameters and by a representation of the zero-order hold (Section F.7.2) and a delay T_d to account for processing. The total delay is represented approximately by $\exp[-(T_{ref}/2 + T_d)s]$, and if the delay is small compared to the reciprocal of the loop bandwidth f_L, it can be ignored, as in Fig. 9.8. [One would expect that the delay due to computations would not be longer than the time between computations (although pipelining may allow it to be longer), in which case $T_{ref}/2 + T_d < 1.5T_{ref}$.] There is a rule of thumb that the unity gain bandwidth f_L should not exceed $F_{ref}/10$ (Section 8.3). This is related to the 18° of phase shift that would then occur at f_L due to the delay $T_{ref}/2$. If T_d is significant,

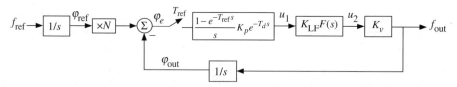

FIGURE 9.7   Mathematical diagram of the ADPLL synthesizer.

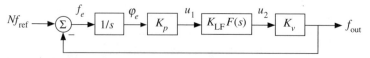

FIGURE 9.8   Simplified mathematical diagram for small delay.

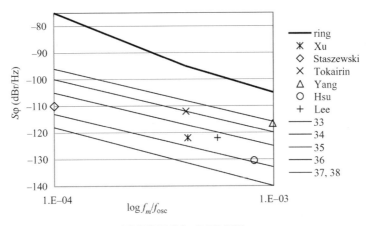

FIGURE 9.9   DCO PSD.

the factor 10 in the rule of thumb should be increased proportionally to the increase in total delay.

Figure 9.8 is a standard PLL diagram like Fig. 1.1*b*, but without the 1/*N*. Thus, the loop gain does not change as F_{out}, and therefore *N*, varies. Note that $u_1, u_2, K_p, K_{LF},$ and K_v are numbers (dimensionless).

9.2.4 DCO

The DCOs are LC oscillators, tuned by switching varactors between high and low capacity states. Figure 9.9 shows points on the output PSD curves of synthesizers reported in some of the sources that are referenced in this section. Many of these synthesizers are first-order loops, so the PSD in these −6 dB/octave regions could be due to attenuated internal noise, but we know that the oscillators are at least this clean. They can be compared to PSDs from LC oscillators in Fig. O.1, which are represented by the lines in Fig. 9.9 and which are numbered as they are in Fig. O.1. We see that the ADPLL synthesizer PSDs fall in the midst of the recorded oscillator PSDs. ICs often use ring (delay-line) oscillators, which are like their digital circuitry. The phase noise of some CMOS ring oscillators is also shown (from Fig. F.3.38*e*) and can be seen to be much higher. Significant space is sacrificed for spectral purity in the use of LC oscillators, which are relatively huge.

Fine tuning is commonly achieved by modulating the switching voltage to a small varactor with a controlled duty factor, often using ΣΔ modulation to spread the resulting noise. The use of ΣΔ modulation to spread the noise from the switched tuning is somewhat surprising, since this tends to move the noise to high frequencies and the loop does not suppress noise introduced to the DCO. Staszewski and Balsara [2006, p. 126] show that the resulting noise does rise above the DCO noise at high offsets, whereas white dither would parallel the DCO noise in the −6 dB/octave region. Nevertheless, their resulting noise was below the limit specified by a mask from a

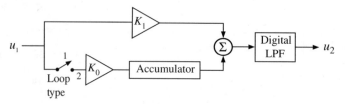

FIGURE 9.10 Loop filter.

GSM specification. While they show this noise as a density, Xu et al. [2009] show individual spurs, which are below -57 dBc. Both used second-order $\Sigma\Delta$ modulators, with noise peaking between 80 and 400 kHz offset.

Some synthesizers use DACs for fine tuning (which would seem to be less compatible with the concept of *ADPLL*).

9.2.5 Loop Filter

The loop filter is shown in Fig. 9.10. Inclusion of the accumulator makes the loop type-2 (Section F.6), putting a second pole at the origin. Unlike the analog case, where the imperfect integrator makes the second pole at the origin only an approximation, the accumulator is a true integrator. It does not run out of gain at low frequencies, although its dynamic range is still finite.

9.2.6 Synchronization

There is a tendency for the DCO to be pulled by digital signals that become coupled into the oscillator circuitry. Pulling can change the frequency of an open-loop oscillator and can cause phase modulation of a locked oscillator. Staszewski and Balsara take great care to accurately resynchronize the transition of F_{ref} with a transition of the DCO, creating a clock that has the same average frequency as the reference signal but which transitions in synchronism with F_{out}. Thus, the reference signal in Fig. 9.6 is actually this resynchronized reference in their synthesizers. They use this resynchronized reference throughout the synthesizer. If such a signal leaks into the oscillator circuitry, it may cause some phase shift, but since the time between the transitions of the DCO and of the leaked signal are constant, it will not modulate the phase of the DCO.

9.2.7 Phase Noise

9.2.7.1 *In-Band Noise, Critical Source* Any number has finite resolution. Finite resolutions (i.e., round-off or truncation errors) are sources of noise throughout the loop. They can usually be made small enough to be inconsequential by the use of enough digits to represent any quantity to a sufficient accuracy, but one place where this is a problem is in the measurement of φ_e. If we just subtract the state of the

frequency counter from the desired phase, the resolution will be $\pm \Delta t/2$, where $\Delta t = T_{\text{out}}$. Assuming a uniform probability in the actual time difference within these limits, the variance of the time error is

$$\sigma_t^2 = (\Delta t)^2 / 12. \tag{9.15}$$

The corresponding phase variance referenced to f_{ref} is

$$\sigma_{\varphi,\text{ref}}^2 = \frac{1}{12} \left(\frac{\Delta t}{T_{\text{ref}}} \text{cycle} \right)^2, \tag{9.16}$$

so the SSB PSD so referenced is[55]

$$\mathcal{L}_{\text{ref}} = S_{2,\varphi,\text{ref}} = \frac{\sigma_{\varphi,\text{ref}}^2}{F_{\text{ref}}} = \frac{1}{12 F_{\text{ref}}} \left(\frac{\Delta t}{T_{\text{ref}}} 2\pi \, \text{rad} \right)^2 = \frac{(2\pi)^2}{12} \frac{\Delta t^2 F_{\text{ref}}}{\text{cycle}^2}. \tag{9.17}$$

If we compare this to Eq. (4.6) (with $I_{\text{cp}} \gg I_{\text{knee}}$), we obtain

$$k_1 = \frac{(2\pi)^2}{12} \frac{\Delta t^2}{\text{cycle}-\text{sec}}. \tag{9.18}$$

For the value of k_1 used for our noise analysis in Section 4.1.6, $\Delta t = 25$ ps, which is the period at 40 GHz. Thus, we require some means of reducing the range of indeterminacy below 1 cycle if we are to operate at lower frequencies. Failure to do so would cause high levels of in-band phase noise. Great ingenuity has been applied to the reduction of Δt.

We should note that Eq. (9.15) is widely used, but the assumed probability distribution often depends on the taking of samples from well-separated or essentially independent Δt segments, in order that we can assume randomness in the actual value that occurs each sample. This requirement is not always well met. We will discuss its applicability further in Section 9.2.13.

9.2.7.2 *Improving Resolution*

A common method for improving the resolution of the phase error is to obtain a fine measurement of the time of the reference transition by using inverter delays as a measurement increment in a time-to-digital converter (TDC). Figure 9.11 illustrates a TDC [Staszewski and Balsara, 2006]. The DCO output passes through a series of inverter delays, each of which in connected to the D input of a bistable. The bistables are all set to the values at their inputs when the reference signal rises. Waveforms are illustrated in Fig. 9.12. The outputs from the bistables can form a word, as illustrated along the vertical line. In this case, the word is 001111000. The first *two* zeros indicate *two* delays between the resetting of the DCO signal and the rise of the reference. These zeros plus the next four ones show that the reference transition occurred between *six* and *seven* delays after the rising edge of the DCO output. Either fact can be used to obtain the phase within $\pm$ half

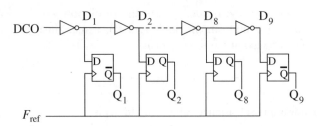

FIGURE 9.11   TDC concept.

an inverter delay, once the delays have been calibrated, that is, related to the period of the DCO signal. Calibration can be based on the number of delays that occur during the DCO period, averaged over many periods (Section A.2.3).

Staszewski and Balsara obtained 30 ps inverter delays with a 130 nm CMOS process. From Eq. (9.18), this corresponds to $k_1 = 3 \times 10^{-21}$/Hz, which is equivalent to -205 dBc/Hz for Banerjee's white noise factor (Section 4.1.3) and is just 2 dB higher than the level that we used for our noise model in Section 4.1.6. It is within the range of -203.5 to -220 dBc/Hz reported for National Semiconductor synthesizer ICs in Table 14.1 of Banerjee [2006], although near the upper end. Based on Eq. (4.6), multiplied by F_{out}/F_{ref}, it provides an in-band noise level of -90 dBc/Hz with their 2 GHz F_{out} and 13 MHz F_{ref}.

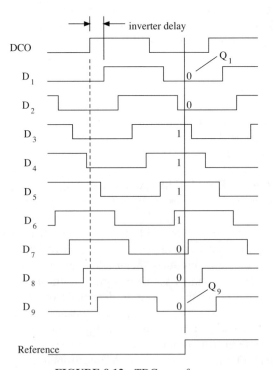

FIGURE 9.12   TDC waveforms.

These resolutions are improving as inverter delays shrink with feature sizes over time, but researchers have not been content to wait.

Tokairin et al. [2010] used two delay chains (at 2.1–2.8 GHz) with slightly different unit delays to obtain measurement resolution well below the inverter delays. Simulations indicated $\Delta t = 5$ ps. Measured noise implied 8 ps, equivalent to $k_1 = 2 \times 10^{-22}$/Hz $\Rightarrow -217$ dBc/Hz. Temporiti et al. [2009] got similar measured noise using a "vernier delay line."

Hsu et al. [2008] used a ring oscillator, gated on during the period measured, to achieve 6 ps resolution, but an improvement was expected due to noise shaping that occurs with their circuit. They achieved effectively $k_1 = 5.8 \times 10^{-23} \Rightarrow -222$ dBc/Hz, equivalent to $\Delta t = 4$ ps.

Lee et al. [2009] used a time amplifier circuit, which increases the time difference between two events, to improve the resolution of the TDC. Based on the achieved noise level, they obtained an effective resolution of 1.1 ps, equivalent to $k_1 = 4 \times 10^{-24} \Rightarrow -234$ dBc/Hz, 14 dB better than the best value given by Banerjee above.

Yang et al. [2010] achieved $k_1 \Rightarrow -204$ dBc/Hz, operating their loop, with integer N, as a bang-bang servo; that is, they did not use a TDC and, once their loop had settled, the PD sensed only whether the DCO phase was early or late. See more references for TDCs in Staszewski et al. [2009].

9.2.8 Reference Spurs

The ADPLL synthesizer is subject to modulation of the DCO by noise from digital signals, as may be expected when an analog circuit (which the DCO is, internally) exists in proximity to digital signals. The four researchers just referenced reported spur levels related to F_{ref} between -65 and -40 dBc. (See Section 4.4.2 for such levels in $\Sigma\Delta$ synthesizers.)

9.2.9 Fractional Spurs

Ideally, the ADPLL synthesizer can produce fractional frequencies without producing phase modulation. Each reference period, the ideal phase from the accumulator increases by $N = N_{int} + n_{fract}$ and the cycles counted by the frequency counter, plus the fraction of a cycle measured by the TDC circuit, increase by the same amount, producing a constant phase error. However, this depends on the resolution and accuracy (linearity) of the TDC.

When $n_{fract} = 0$, the value from the TDC will be the same every cycle, so linearity is not important. But when there is a fractional value, the TDC output will change each cycle, and if the TDC is not linear, φ_e will not be constant and a signal will be produced to modulate the DCO (see Section 4.4.1 regarding the similar problem in $\Sigma\Delta$ synthesizers). Linearity requires that each inverter have the same delay and that the delay be properly calibrated (Section A.2.3) with respect to T_{out}.

Even if the TDC circuit were perfectly accurate, however, modulation would be produced unless its resolution matched n_{fract}. For a simple example [Temporiti et al.,

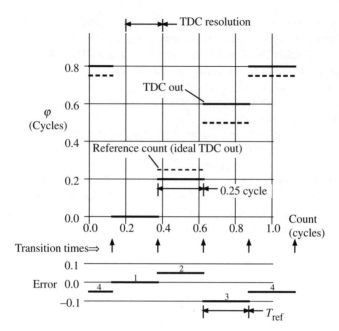

FIGURE 9.13   TDC resolution and fractional spurs. The x-axis represents time, but fractions of the frequency count are also shown.

2009], if n_{fract} should equal 0.25, the measured phase would move across 1 output cycle in four reference periods (Fig. 9.13). If the TDC should provide five perfectly spaced incremental subdivisions of T_{out}, the TDC would not produce five equally spaced numerical values. Only four of the available five levels would be used and an error waveform would be produced that would repeat with a period of $4T_{\text{ref}}$, causing phase modulation and sidebands spaced at multiples of $f_{\text{fract}} = 0.25 \, F_{\text{ref}}$ from center. Moreover, since the average error in this sequence is between 0.1 and 0.5 cycle (depending on the phase error), attaining a required error value (e.g., 0 if the loop were type-2) could require phase modulation that periodically changed the integer error.

For $n_{\text{fract}} = 1/16$ or $n_{\text{fract}} = 5/16$, $16T_{\text{ref}}$ would be the repetition interval, producing spurs offset by multiples of $F_{\text{ref}}/16$. But if the TDC had 16 increments, there would be no error waveform, assuming perfect accuracy. However, if the TDC increments were not uniform, there would be modulation at multiples of $F_{\text{ref}}/16$. Of course, the maximum error becomes smaller with finer resolution.

Fractional spur levels reported by four researchers[56] varied between −42 and −61 dBc except that Tokairin et al. reported none observed. Test conditions, such as the fractional frequencies tested and filter attenuation, vary and are not always specified completely. However, Hsu et al. show these in some detail. There we can see how larger fractional spurs occur at small fractional frequencies (n_{fract} close to 0 or 1) within the loop filter bandwidth. Temoriti et al. show how spur levels changed from −45 dBc with their TDC mismatch correction circuit in place to −34 dBc without it and to −27 dBc when they purposely add 2.5% to the ideal delay of each element. These levels were for harmonics of f_{fract} within the loop bandwidth at 3 GHz F_{out}. Weltin-Wu et al. [2010] employed TDC dithering plus feedforward compensation to reduce fractional spurs from as high as -38 dBc to no more than -58 dBc with only 1 dB increase in broadband noise, at 3.5 GHz. Again, these spurs were within the loop bandwidth.

Although it is difficult to associate the various TDC circuits with particular fractional spur levels, some of the levels reported alert us to the importance of such spurs in ADPLL synthesizers.

9.2.10 Modulation Response

Vamvakos et al. [2006] injected white noise into an ADPLL to observe the effect on the spectrum. They found that the resulting errors in indicated phase were essentially independent of the actual phase with $n_{\text{fract}} = 0.01$, but were both greatest and least when $n_{\text{fract}} = 0$, being greatest when the phase occurred near a TDC boundary. They also injected sinusoidal noise and found that the inherently nonlinear nature of the TDC causes sidebands to be produced closer to the carrier than the injected frequency. This is somewhat reminiscent of similar effects with $\Sigma\Delta$ modulators and is of concern because the loop filter is less effective close to the carrier. Staszewski et al. [2009] found that operating an ADPLL with $n_{\text{fract}} \approx 0$ tends to produce close-in phase noise during modulation due to poor randomization [violation of the assumptions required for Eq. (9.15)] and that the problem at these frequencies could be reduced by jittering f_{ref}, at some cost to the noise in other channels (values of n_{fract}).

9.2.11 $\Sigma\Delta$ Cancellation

As we have seen, the ADPLL synthesizer can ideally generate frequencies that are not integer multiples of F_{ref} by just including fractions in the value of N that is updated at a rate of F_{ref} (Fig. 9.6). However, Hsu et al. do not include the fraction in N, but instead modulate the divider with $\Sigma\Delta$ modulation. They then cancel the resulting noise in u_1, using a "digital correlation loop" to adjust the cancellation gain. This cancellation should be accurate because this and also the modulation that is being canceled are numbers, not analog signals. However, the process still depends on the linearity of the measured phase error and, therefore, the TDC. The cancellation reduced out-of-band noise as much as 16 dB.

9.2.12 Simulation

Although we have not modeled an ADPLL in Simulink, we anticipate its usefulness as an aid to better understand the ADPLL and as a platform for experiments such as those done with $\Sigma\Delta$ synthesizers.

The book by Staszewski and Balsara [2006] is an unusually detailed reference on the design of ADPLL synthesizers. It includes discussion of simulation using VHDL, which the authors hold to be superior to other methods. They devoted considerable effort to generating appropriate noise sources for simulating the DCO (see also

Staszewski et al. [2005]). Their model closely approximates their IC. This is important in IC development to maximize the probability of a successful design before the expensive fabrication is begun.

Our Simulink models, on the other hand, have not been detailed representations of actual circuits. We have used them as learning tools to study the effects of $\Sigma\Delta$ modulators in synthesizers. We did not include multiple noise sources, but concentrated on the noise created by the $\Sigma\Delta$ process. This allowed us to observe the effects that we were trying to understand, effects that might have been hidden by other noise sources. We analyzed typical designs in Chapter 4 without creating new models. Rather, we combined what we had learned about noise due to $\Sigma\Delta$ modulation with our knowledge of other noise sources.

Simulink models of the ADPLL are beyond the scope of this book, but we show an initial plan for such a model in Section A.2.4. Modeling of the ADPLL synthesizer with a high-level language is discussed by Syllaios et al. [2008].

9.2.13 Dead Zone

Let us consider further the applicability of Eq. (9.15) to describing the effects of finite phase resolution. When a fractional frequency is synthesized, the positive transition of the reference (Fig. 9.12) ordinarily occurs sequentially at various positions within various delay increments, supporting the characterization of the error due to the finite resolution as being evenly distributed over a delay increment, as per Eq. (9.15). On the other hand, Section 9.2.10 would seem to suggest that the applicability of Eq. (9.15) is not supported very well when n_{fract} is 0 or very small.

When F_{out} is an integer multiple of F_{ref}, the reference transition would ideally fall in the same increment each T_{ref}, but the value of the sampled DCO phase depends on the phase error that is required to tune the DCO (in the case of a type-1 loop) and on any value in the reference accumulator (Fig. 9.6) when it began to be incremented by whole numbers (if it was not first reset). This phase generally will not have a fractional part that equals any of the phase increments in the TDC. Therefore, required phase error will only be attained on an average, as the TDC output changes between adjacent values in a pattern that produces the required average, and that would produce some randomness in the error.

There can be conditions under which the reference transition occurs repeatedly in the same delay increment. One likely set of conditions that would cause repeated sampling in the same delay increment is the type-2 loop with a reference accumulator that is reset prior to being driven by an integer. The required phase error is then zero and, since the phases being compared are integers, the required TDC output is zero. Thus, the reference transition would tend to occur repeatedly in the delay increment corresponding to zero output. This would seemingly be a dead zone, analogous to the dead zone experienced in phase frequency detectors (Section F.5.7.2), and, as in that case, output phase noise can be expected to "blossom" because the small phase deviations will not be seen. They will not affect the detected phase; the loop is essentially open. (The sampled phase may also spend significant time within such a dead zone if $n_{fract} \neq 0$ but $n_{fract} \approx 0$.)

Although some noise degradation has been reported when $n_{\text{fract}} \approx 0$, we would expect a dead zone to produce a loss of the normal attenuation, due to feedback, of the DCO's inherent phase noise, but such an effect does not seem to have been observed and reported. Jittering (Section 9.2.10) can alleviate the problem, much as adding noise to the signal driving a midtread ADC can cause the signal to appear in the digitized spectrum, even though it is too small to drive the ADC by itself. Signals that are too small to be seen without the jitter can still influence the result when jitter causes a change in the output. If the jitter produces phase error outputs at a rate that is high compared to the loop bandwidth, it could preserve the feedback that attenuates the DCO noise. But what may be the source of such jitter in the absence of intentional modulation? One possibility is reference noise (Section A.2.2).

Without experimental data or a model to simulate these effects, we have no way to verify them or to determine under what conditions they may be important. Again, we see the value of a model, this time in its absence.

APPENDIX A

ALL DIGITAL

A.1 FLYING ADDER CIRCUITS

Mair and Xiu [2000] show versions of the Flying Adder that provide for controllable phase delay and for higher speed operation. A version of the latter circuit [Xiu and You, 2002] is shown in Fig. A.1. The objective here is not only to increase maximum output frequency but also to avoid a potential problem when addresses to the mux change, at which time some intermediate state may occur during switching and erroneously trigger the output bistable. However, it would seem that the output flip-flop could be made to ignore these glitches if an appropriate delay were inserted into the feedback from output to D input (Fig. A.2) or possibly by other logic design methods.

In Fig. A.1, the input to the top register receives the sum of the MSBs from the accumulator and $\lfloor K/2 \rfloor$, that is the $(m-1)$ MSBs from k, shifted down 1 bit. They will select a phase that is midway between outputs from the lower mux, if K is even, otherwise a little less. The switch S_{out} selects the two muxes alternately, so each output from the lower mux will be followed by a delayed output from the upper mux. As a result, the signal at f_x will be square (if K is even) and can provide a uniform pulse at frequency f_x. The mux to which flip-flop FF is connected will not receive an address change while it is connected. When a positive edge drives FF, the mux will be switched, but S_{out} will connect FF to the other mux at that time. An extra register has been inserted into the address paths to give the Adder more time (pipelining), thus increasing the maximum speed.

Advanced Frequency Synthesis by Phase Lock, First Edition. William F. Egan.
© 2011 John Wiley & Sons, Inc. Published 2011 by John Wiley & Sons, Inc.

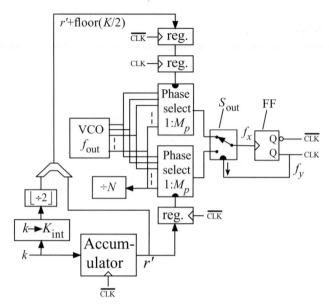

FIGURE A.1   Improved Flying Adder circuit.

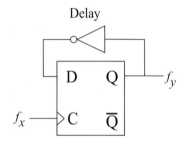

FIGURE A.2   Divide-by-two desensitized to multiple triggers.

Xiu and You also show how to use four muxes to further increase the speed.

A.2   ADPLL SYNTHESIZER

A.2.1   Alternative Architecture

If we follow the frequency counter in Fig. 9.6 by a digital differentiator (transfer function $1 - z^{-1}$) and follow that by an accumulator (a digital integrator, with transfer function $1/(1 - z^{-1})$), the added blocks have a net transfer function of 1, so the loop is not fundamentally changed. Now, however, there is an accumulator at both inputs to the summer, so they can both be moved through the summer, creating the block diagram shown in Fig. A.3. This is an alternative architecture that, according to Staszewski and Balsara [2006], has advantages.

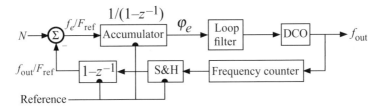

FIGURE A.3   Alternative architecture.

The output of the frequency counter here includes the fractional values from the TDC, as it did in Fig. 9.6, and these are carried through the subsequent blocks. The lower input to the summer is the change in count (phase in cycles) that occurred during T_{ref}, which equals

$$f_{out}T_{ref} = \frac{f_{out}}{F_{ref}} \text{cycle.} \tag{A.1}$$

This value is subtracted from N to give the normalized frequency error,

$$N - \frac{f_{out}}{F_{ref}} = \frac{F_{out,desired} - f_{out,actual}}{F_{ref}} = \frac{f_e}{F_{ref}}, \tag{A.2}$$

which is integrated in the accumulator, at a clock rate of F_{ref}, to produce the phase error φ_e.

A.2.2 Reference Jitter and the Dead Zone

We would like to see if the noise floor accompanying the reference oscillator signal can cause enough jitter, when this relatively slow signal is converted to a digital clock, to move the phase from one TDC segment to another. If it does so often enough, these events may act like a sampling clock and permit effective feedback at frequencies within the loop bandwidth f_L.

If a phase is characterized by a constant mean value and a PPSD S_φ, the variance of the change in phase between two measurements separated by T_x is given by

$$\sigma_\varphi^2 = \langle [\varphi(t+T_x) - \varphi(t)]^2 \rangle = 4 \int_0^{f_{max}} S_\varphi \sin^2 \left(\frac{f_n T_x}{2} \right) df_n. \tag{F.11.45}\,\text{(A.3)}$$

Let us simplify the integration by starting it at a peak of the sine, $f_n = \text{cycle}/(2T_x)$, and using the average value of $\sin^2$, 1/2, at higher frequencies. Then,

$$\sigma_\varphi^2 \approx 2 \int_{\text{cycle}/(2T_x)}^{f_{max}} S_\varphi df_n. \tag{A.4}$$

Roughly speaking, starting at cycle/$(2T_x)$ eliminates frequencies that are too low to change much in a time T_x. For white noise, the integral becomes

$$\sigma_\varphi^2 \approx 2S_{\varphi,\text{flat}}\Delta f, \qquad (A.5)$$

where

$$\Delta f = f_{\text{max}} - \text{cycle}/(2T_x). \qquad (A.6)$$

However, since we are interested in jitter between the reference and the DCO, the S_φ of interest is the PPSD of the phase error, so we must also multiply by $1 - H(f_m)$, assuming that the feedback is effective. If we then ignore the attenuated region of S_φ below f_L, Eq. (A.6) becomes

$$\Delta f = f_{\text{max}} - \text{maximum}(\text{cycle}/(2T_x), f_L). \qquad (A.7)$$

We are interested in measurement times that satisfy

$$T_x \ll \frac{\text{cycle}}{f_L} \qquad (A.8)$$

to see whether the phase may change TDC increments at a frequency much faster than the loop bandwidth, and thus provide a sampling process that would support the suppression of DCO phase noise out to that frequency. If $f_{\text{max}} \gg f_L$, which we would expect, we can begin integration at cycle/$(2T_x) \gg f_L/2$ [also implying cycle/$(2T_x) > f_L$ in Eq. (A.7)] in accordance with (A.8) and still have $f_{\text{max}} \gg$ cycle/$(2T_x)$ in Eq. (A.6), allowing the approximation

$$\Delta f \approx f_{\text{max}}. \qquad (A.9)$$

The corresponding time jitter at a frequency F_{ref} can be obtained from

$$\Delta t = \Delta\varphi/F_{\text{ref}} \Rightarrow \sigma_t = \sigma_\varphi/F_{\text{ref}}, \qquad (A.10)$$

so the variance of the jitter would be

$$\sigma_t^2 \approx 2S_{\varphi,\text{flat}}\left(\frac{\text{cycle}}{2\pi\,\text{rad}}\right)^2 \frac{f_{\text{max}}}{F_{\text{ref}}^2}. \qquad (A.11)$$

Here, we have combined Eqs. (A.5), (A.9) and (A.10). Suppose the reference oscillator has $F_{\text{ref}} = 10\,\text{MHz}$, a noise floor of $-130\,\text{dBr/Hz}$ (as assumed by Staszewski and Balsara, in their simulation, for an inexpensive crystal oscillator[A1]), and the floor extends to $f_m = f_{max} = 10$ MHz. Then, Eq. (A.11) gives $\sigma_t \approx 23$ ps. So we can expect this standard deviation in the phase change over any period T_x, which must be long compared to cycle/f_{max} and short compared to cycle/f_L. Since the latter is much greater than the former, we can apply this expectation many times during the period of f_L; so, if this value of σ_t is appreciable compared to a TDC increment, we would expect jitter at a high rate compared to f_L.

If the noise floor were due to circuit noise, deep submicron transistors may respond out to 100 GHz.[A2] In this case, Δf would be 10^4 higher and we could obtain the same σ_t with $\mathcal{L}_\varphi \approx -170$ dBc/Hz,[A3] which corresponds to the thermal noise for a -7 dBm oscillator signal (e.g., 1 V peak–peak in 600 Ω). The two effects together could double σ_t^2.

In summary, it is not unlikely that the noise floor that accompanies the reference signal, where it is converted to a digital clock, would cause enough jitter to produce φ_e changes often enough to enable the desired feedback, even when the phase is in the "dead zone."

A.2.3 Reference Jitter and Calibration

Reference jitter can also improve the accuracy of the TDC calibration by permitting the number of inverter delays that correspond to T_{out} to be averaged over measurements made with different time differences T_e between DCO and reference transitions. Without this averaging, T_{out} (actually $T_{out}/2$ in the method of Staszewski and Balsara [2006]) could only be obtained in terms of whole numbers of inverter delays. Section 9.2.13 describes conditions under which T_e does not vary in the absence of jitter (e.g., a dead zone).

The TDC waveforms of Fig. 9.12 are shown again in Fig. A.4, but here rather than showing one reference waveform rising at one time, we have shown regions where the reference may rise. TDC calibration, according to the method described by Staszewski and Balsara, embodies counting the number of delayed DCO outputs that are high at the time the reference rises. This indicates how many delays fit into $T_{out}/2$. In Fig. 9.12, the number is 4, but that can change if the time of occurrence of the rise in the reference waveform changes. We can see this at the bottom of Fig. A.4, where the number of Q outputs, from the bistables in Fig. 9.11, that are 1 is plotted against the time of the reference rise. Sometimes the number is 4 and sometimes it is 5, but the average, over an integer number of inverter delays, is the true width of $T_{out}/2$ expressed in delays. Staszewski and Balsara averaged 128 measurements to obtain 1 ps accuracy in the calibrated inverter delay (inverter delay ≈ 30 ps, $T_{out}/2 \approx 200$ ps). Without T_e variation, each measurement would have been made at the same time, relative to the DCO waveform, and would have given the same, integer, answer (4 or 5 in the example), so variation is essential for improving the calibration accuracy.

Again, we see the importance of reference jitter (phase noise), whether natural or induced. The important noise has frequencies higher than f_L, so it tends not to be followed by φ_{out}. Of course, it will often be accompanied by lower frequency noise that will be followed at the output.

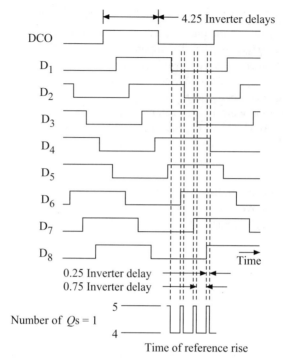

FIGURE A.4 TDC calibration.

A.2.4 Initial Plan for a Model of an ADPLL Synthesizer

Modeling the ADPLL synthesizer is beyond the scope of this book. (At some point, one must stop experimenting and start writing!) However, we will consider an initial plan for a simple model, one that would allow us to observe some of the effects of the TDC. It would probably undergo revisions and, eventually, further development, just as did the models described in Chapter 6. This proposed model will avoid simulation of the output sinusoid, which we expect would make it faster (Section 6.14).

The output of the model in Fig. A.5 is the frequency f_{out} that is integrated to produce the output phase. We use a Simulink integrator that can be reset to 0 when it reaches N_c cycles, the capacity of the frequency counter (Fig. 9.6). Then, we multiply the phase by n_i, the number of increments in the TDC, and process the result through a floor function that drops the fractional part. We then divide by n_i, giving a numerical value corresponding to the whole cycle count plus the number from the TDC. At this point, we have simulated the information available from a perfectly calibrated TDC.

The divide ratio N is fed into an accumulator of capacity N_c, just as in Fig. 9.6. This can be a modification of the first accumulator used in the $\Sigma\Delta$ models, wherein, however, we would make it roll over at N_c rather than at 1. It should retain the ability to be initialized to a specified value. At the (positive) transition of F_{ref} (which we may

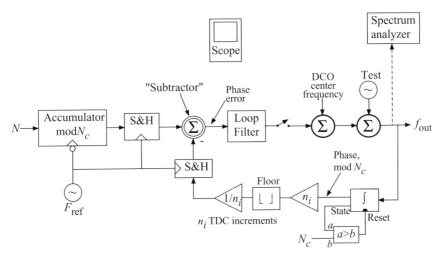

FIGURE A.5 Initial plan for a Simulink model of an ADPLL synthesizer.

want to square up), both phase values are sampled. (We are not simulating resynchronization of F_{ref} as per Section 9.2.6.) The values are subtracted in the "subtractor," which is detailed in Fig. A.6, to produce the phase error that can be positive or negative. This is processed through a loop filter, as was done in our $\Sigma\Delta$ models. With the switch closed, the filter output is added to a number giving the DCO center frequency, producing f_{out}, and completing the loop.

The subtractor changes the $\text{mod}N_c$ error value e_2 into a signed number at e_5, the first bit becoming the sign. Values of e_2 that are higher than $N_c/2$ then become two-complement negative numbers. See the examples in the table in Fig. A.6.

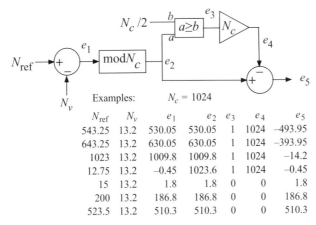

Examples:		$N_c = 1024$				
N_{ref}	N_v	e_1	e_2	e_3	e_4	e_5
543.25	13.2	530.05	530.05	1	1024	−493.95
643.25	13.2	630.05	630.05	1	1024	−393.95
1023	13.2	1009.8	1009.8	1	1024	−14.2
12.75	13.2	−0.45	1023.6	1	1024	−0.45
15	13.2	1.8	1.8	0	0	1.8
200	13.2	186.8	186.8	0	0	186.8
523.5	13.2	510.3	510.3	0	0	510.3

FIGURE A.6 Subtractor for Fig. A.5.

We can add a small sinusoid (test) to the frequency to represent the noise modulation at a particular value of f_m in order to verify that it is multiplied by the error transfer function, $1 - H(f_m)$ (comparing f_{out} with the switch opened and closed), or we can use test to represent out-of-band frequency (phase) modulation to observe what noise it may produce (Section 9.2.10).

A scope can be used to observe the various waveforms, as we did in the $\Sigma\Delta$ simulations, and a spectrum analyzer can show the spectrum of f_{out}, which can be converted to phase by dividing by f_m (Sections 6.1.7 and S.7).

At this level, the primary effect that can be observed experimentally is the stepped nature of the TDC transfer function under various conditions (integer-N, fractional-N, and type-1 or type-2 loop), but the model is likely to grow in complexity as we continue our experiments.[A4] We may add white noise at the subtractor as done in Egan [1998, Section 18.M] or Egan [2007, Section i.17.M] to observe its effect under various conditions.

Syllaios et al. [2008] provide pseudocode for an accurate simulation of the ADPLL synthesizer. As in the model discussed here, it concentrates on computing the critical times rather than computing RF waveforms.

APPENDIX C

FRACTIONAL CANCELLATION

In this appendix, we show how the accumulator contents in a simple (composed of first-order stages) MASH modulator are related to the phase error and can, therefore, be used to compensate it. Numbers will be normalized to the capacity of the accumulator, which will, therefore, equal 1.

C.1 MODULATOR DETAILS

Let us consider the implementation of the modulator in Fig. 2.1 in more detail. Figure C.1 shows the general concept of a first-order MASH modulator. The input is a binary fraction. An accumulator drives a quantizer and the quantizer output is fed back to subtract 1 from the input. A possible implementation is shown in Fig. C.2. Here, only 4 bits are shown for simplicity in the drawing. Many more bits are usual in practice, but the concepts remain the same. There is a storage register to store the previous output and probably also an input register, although it is sometimes simpler to consider the realization without one, as is done in Section F.8.3.5. Here, we will concentrate on the accumulator with an input register.

Shortly after the clock causes the registers to be updated, the arithmetic sum of the input and the previous output appear at the output. The n (=4) least significant bits at the output are sent to the storage register. The most significant bit becomes the output of the quantizer, Q_{out}. The process of subtracting 1 from the input occurs automatically as the n-bit number at a rolls over (i.e., the sum becomes greater than the capacity

Advanced Frequency Synthesis by Phase Lock, First Edition. William F. Egan.
© 2011 John Wiley & Sons, Inc. Published 2011 by John Wiley & Sons, Inc.

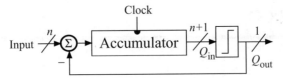

FIGURE C.1   Modulator with a single quantizer.

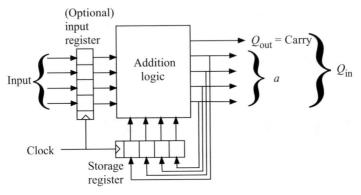

FIGURE C.2   A possible implementation of the modulator in Fig. C.1.

of the register so the MSB is lost, leaving a smaller number). Because Q_{out} is not fed back to the storage register, it appears only when a rolls over. The mathematical diagram is shown in Fig. C.3a and, in a more compact form, in Fig. C.3b. The quantizer is represented as a summer wherein the quantizing noise q is added to its input Q_{in}. The characteristic of q is such that it equals the negative of Q_{in} until Q_{in} equals or exceeds 1, at which time its value increases (becomes less negative) by 1, causing Q_{out} to become 1.

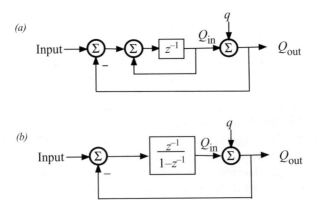

FIGURE C.3   Mathematical diagram of modulator with input register. Diagram in (a) is simplified in (b).

From control theory, we can write

$$Q_{out} = [\text{input}]\frac{z^{-1}/(1-z^{-1})}{1+(z^{-1}/(1-z^{-1}))} + q\frac{1}{1+(z^{-1}/(1-z^{-1}))} = [\text{input}]z^{-1} + q(1-z^{-1}).$$

(C.1)

From Fig. C.2, we see that

$$Q_{in} = a + Q_{out}$$

(C.2)

and from Fig. C.3b,

$$Q_{in} + q = Q_{out},$$

(C.3)

showing that

$$q = -a.$$

(C.4)

In other words, the quantization noise is the negative of the contents of the accumulator register (excluding the carry bit).

C.2 FIRST ACCUMULATOR

For the first MASH accumulator, $[\text{input}] = n_{fract}$, a constant at any synthesized frequency; so the output [Eq. (C.1)] consists of the desired n_{fract} and a term embodying the variations in the output, $-a(1 - z^{-1})$, which is noise. The variation of phase at the PD equals the variation in the modulator output divided by $\overline{N}(1-z^{-1})$ [Eq. (Q.6)], that is, $-a/\overline{N}$, and the voltage produced thereby equals this times K_p cycle. For this reason, the contents of the accumulator register, scaled by $K_p$cycle$/\overline{N}$, provide proper compensation for the waveform at the PD output.

C.3 SECOND ACCUMULATOR

When a second accumulator is provided in a second-order MASH (Fig. 2.10), its input is $-q$ ($=a$). The transfer function for its input is again z^{-1}. This is multiplied by $(1 - z^{-1})$ and added to the (delayed) output from the first stage, canceling the quantization noise in Eq. (C.1) (while n_{fract} continues to be applied to modify N). The other component in the output of this second stage is $q_2(1-z^{-1})$, where q_2 is the quantization noise from the second stage. Thus, the quantization noise from the first stage is replaced by quantization noise from the second stage. Just as the contents of the accumulator register, multiplied by $K_p$cycle$/\overline{N}$, provided compensation for q from

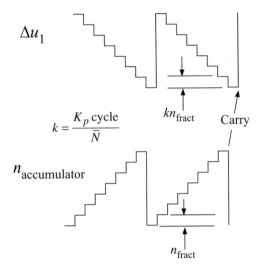

FIGURE C.4   PD voltage and accumulator content.

the first stage, so do the contents of the second accumulator's register a_2, multiplied by $K_p\text{cycle}/\overline{N}$, provided compensation for q_2 from the second stage. However, since the output of the second stage is multiplied by $(1 - z^{-1})$ before affecting N, a_2 must also be multiplied by $(1 - z^{-1})$ to preserve the relationship between the quantization noise being canceled and the canceling waveform.

The accumulator overflows at the clock before the division with the new divide number, so timing details must be taken into account for good cancellation of the PD output waveform (see Fig. C.4).

C.4   ADDITIONAL ACCUMULATORS

Each time the order of the MASH circuit increases by 1 (Fig. C.5), the quantization noise from the previous level is replaced by the quantization noise from the new level and the contents of the new accumulator's register become the source of the signal to cancel the quantization noise at the PD, in each case being processed identically with the processing of the output of the last stage.

In equations, the output from the mth MASH level is

$$Q_{\text{out},m} = -q_{m-1}z^{-1} + q_m(1-z^{-1}) \qquad (\text{C.5})$$

and the output from the next higher level is

$$Q_{\text{out},m+1} = -q_m z^{-1} + q_{m+1}(1-z^{-1}). \qquad (\text{C.6})$$

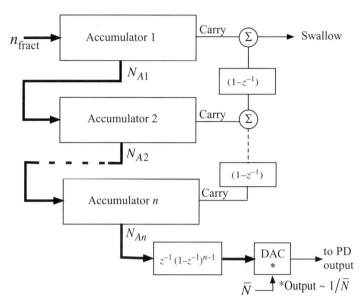

FIGURE C.5 Multiple accumulator stages. Unit delays that are required to properly synchronize the separate carry outputs are not shown.

The output from this upper level is multiplied by $(1-z^{-1})$ before being added to the next lower level, which is delayed by one clock cycle relative to the upper level:

$$Q_{\text{out},m+1}(1-z^{-1}) + z^{-1}Q_{\text{out},m} = \begin{bmatrix} -q_m z^{-1}(1-z^{-1}) + q_{m+1}(1-z^{-1})^2 \\ -q_{m-1}z^{-2} + q_m z^{-1}(1-z^{-1}) \end{bmatrix} \quad (C.7)$$

$$= q_{m+1}(1-z^{-1})^2 - q_{m-1}z^{-2}. \quad (C.8)$$

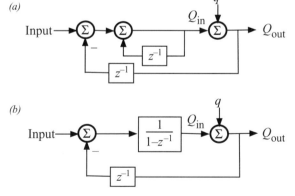

FIGURE C.6 Mathematical diagram of modulator with no input register. Diagram in (*a*) is simplified in (*b*).

Here, we have shown that q_m is canceled due to stage $q_m + 1$, leaving $q_{m+1}(1-z^{-1})^2$ (and also $-q_{m-1}z^{-2}$, but that cancels a similar term in stage $m - 1$), which will arrive at the output N as $q_{m+1}(1-z^{-1})^{m+1}$, assuming there are no higher level stages to cancel it. As before, the order of $(1-z^{-1})^{m+1}$ is reduced by 1 before appearing as PD noise and the register contents a_{m+1} must also be multiplied by $(1-z^{-1})^m$ before being used for cancellation.

C.5 ACCUMULATOR WITHOUT INPUT REGISTER

If the input register in Fig. C.2 is not included, the input appears at the output almost immediately. This tends to simplify our equations, since matching delays are not required to ensure that the outputs are aligned properly. The model for this case is shown in Fig. C.6.

APPENDIX E

EXCESS PPSD

This development follows that of Cassia et al. [2003].

E.1 DEVELOPMENT OF EQ. (2.4)

We first write the Fourier series for the current pulses from the PFD as

$$I(x = n/T) = \sum_{n=-\infty}^{\infty} D_n e^{j2\pi nt/T}, \qquad \text{(E.1)}$$

where

$$x = f_m/\text{cycle} \qquad \text{(E.2)}$$

is the normalized modulation frequency and the amplitude of the nth harmonic is

$$D_n = \frac{1}{T} \int_0^T i(t) e^{-j2\pi nt/T} dt \qquad \text{(E.3)}$$

Advanced Frequency Synthesis by Phase Lock, First Edition. William F. Egan.
© 2011 John Wiley & Sons, Inc. Published 2011 by John Wiley & Sons, Inc.

$$= \frac{I_p}{T} \int_0^T \sum_{m=0}^{L_{sequ}-1} [U(t-mT_{ref}-\Delta t(m))-U(t-mT_{ref})]e^{-2\pi nt/T} dt \tag{E.4}$$

$$= \frac{I_p}{-j2\pi n} \sum_{m=0}^{L_{sequ}-1} e^{-j2\pi nt/T} \Bigg|_{mT_{ref}}^{mT_{ref}+\Delta t(m)}. \tag{E.5}$$

Here $U(t)$ is the unit step, I_p is the magnitude of the current during the PFD pulse, and $T = L_{sequ}T_{ref}$ is the duration of the repeated sequence.

Equation (E.5) represents positive pulses extending from mT_{ref} to $mT_{ref}+|\Delta t(m)|$ when $\Delta t(m) > 0$ and negative pulses extending from mT_{ref} to $mT_{ref}-|\Delta t(m)|$ when $\Delta t(m) < 0$, thus representing the PFD's shift in pulse position as a function of phase. Evaluating (E.5), we obtain

$$D_n = \frac{I_p}{-j2\pi n} \sum_{m=0}^{L_{sequ}-1} e^{-j2\pi nm/L} \left(e^{-2\pi n\Delta t(m)/T} - 1 \right). \tag{E.6}$$

Replacing the last exponential with the first three terms in its expansion, we obtain

$$D_n \approx \frac{I_p}{-j2\pi n} \sum_{m=0}^{L_{sequ}-1} e^{-j2\pi nm/L_{sequ}} \left(1 - 2\pi n\Delta t(m)/T - \frac{1}{2}(2\pi n\Delta t(m)/T)^2 - 1 \right) \tag{E.7}$$

$$= \frac{I_p}{jT} \sum_{m=0}^{L_{sequ}-1} e^{-j2\pi xmT_{ref}} (\Delta t(m) + \pi x\Delta t^2(m)). \tag{E.8}$$

We can also represent Eq. (E.8) as

$$D_n \approx D_{n\_linear} + D_{n\_excess}, \tag{E.9}$$

where the meanings of the individual symbols are apparent by comparing the last two equations.

The Fourier series for the current pulses, $I(x)$ in Eq. (E.1), leads directly to the series for the measured, or indicated, phase error due to quantization, referenced to the output [before multiplication by $H(f_m)$],

$$\Phi_q(s) = \overline{N}I(x)/K_p = \overline{N}I(x)\text{cycle}/I_p, \tag{E.10}$$

with series coefficients derived from Eq. (E.8),

$$D_{\varphi n} = (\text{cycle } \overline{N}/I_p)D_n \tag{E.11}$$

$$\approx (\text{cycle } \overline{N}/I_p)(D_{n\_\text{linear}} + D_{n\_\text{excess}}) \tag{E.12}$$

$$\approx \overline{N}\frac{\text{cycle}}{jT} \sum_{m=0}^{L_{\text{sequ}}-1} e^{-j2\pi xmT_{\text{ref}}}(\Delta t(m) + \pi x \Delta t^2(m)) \tag{E.13}$$

$$= D_{\varphi n\_\text{linear}} + D_{\varphi n\_\text{excess}}. \tag{E.14}$$

We can write the mean square phase (phase power) of the nth spectral line as

$$\frac{|D_{\varphi n}|^2}{2} = \frac{|D_{\varphi n\_\text{linear}}|^2}{2} + \frac{|D_{\varphi n\_\text{excess}}|^2}{2} \tag{E.15}$$

under the assumption that $D_{\varphi n\_\text{linear}}$ and $D_{\varphi n\_\text{excess}}$ are uncorrelated (so there are no cross products). This is based on equal probability of positive and negative $\Delta t(m)$ for all m in the sequence.

From this we derive the two-sided PPSD, the power in a "narrow" frequency band divided by its width,

$$S_{2\varphi q}(f_m) \equiv S_{2\varphi q\_\text{linear}}(f_m) + S_{2\varphi q\_\text{excess}}(f_m), \tag{E.16}$$

by summing the elements of Eq. (E.15) over some number of sequential values of n and dividing by the corresponding bandwidth.

From Eq. (E.13) we see that if

$$S_{2\varphi q\_\text{linear}}(f_m) = K^2 S_{\Delta t}, \tag{E.17}$$

then

$$S_{2\varphi q\_\text{excess}}(f_m) = (\pi x K)^2 S_{\Delta t^2}(f_m). \tag{E.18}$$

This is (Section N.6)[E1]

$$S_{2\varphi q\_\text{excess}}(f_m) = (\pi x K)^2 2 S_{\Delta t}(f_m) * S_{\Delta t}(f_m) \tag{E.19}$$

$$= \left(\frac{\pi x}{K}\right)^2 2 S_{2\varphi q\_\text{linear}}(f_m) * S_{2\varphi q\_\text{linear}}(f_m), \tag{E.20}$$

where $*$ represents convolution[E2] and Eq. (E.17) was used to obtain the last equality. If we also identify $S_{2\varphi q\_\text{linear}}(f_m)$ with Eq. (2.3), this becomes

$$S_{2\varphi q\_\text{excess}}(f_m) = \left(\frac{T_{\text{ref}}}{\overline{N}\,\text{cycle}}\pi x\right)^2 2 S_\varphi(f_m) * S_\varphi(f_m). \tag{E.21}$$

Therefore,

$$S_{2\varphi\_excess}(f_m) = 2\left(\frac{\pi f_m}{\overline{N}\, \text{cycle}\, f_{\text{ref}}}\right)^2 \left\{\frac{(2\pi\,\text{rad})^2}{12 f_{\text{ref}}}\left[2\sin\left(\pi\frac{f_m}{f_{\text{ref}}}\right)\right]^{2(p-1)}\right\}$$

$$* \left\{\frac{(2\pi\,\text{rad})^2}{12 f_{\text{ref}}}\left[2\sin\left(\pi\frac{f_m}{f_{\text{ref}}}\right)\right]^{2(p-1)}\right\} \tag{E.22}$$

$$= \frac{2^{4(p-1)}}{2\overline{N}^2 f_{\text{ref}}^2}\frac{\pi^4}{9}\left(\frac{f_m}{f_{\text{ref}}}\right)^2 \left\{\left[\sin\left(\pi\frac{f_m}{f_{\text{ref}}}\right)\right]^{2(p-1)}\right\} * \left\{\left[\sin\left(\pi\frac{f_m}{f_{\text{ref}}}\right)\right]^{2(p-1)}\right\}\text{rad}^2 \tag{E.23}$$

$$= \frac{k_p}{\overline{N}^2}\left(\frac{f_m}{f_{\text{ref}}}\right)^2 \frac{\text{rad}^2}{f_{\text{ref}}}, \tag{E.24}$$

where

$$k_p = \frac{\pi^4 2^{4(p-1)}}{18}\left\{[\sin(\pi x)]^{2(p-1)}\right\} * \left\{[\sin(\pi x)]^{2(p-1)}\right\}. \tag{E.25}$$

The result of the convolution is shown in Fig. E.1 (detailed calculations are in the spreadsheet Appendix E.xls). Note that this is two-sided PPSD, which equals $\mathcal{L}$.

E.2 APPROXIMATING k_P AS CONSTANT

At low frequencies,

$$\left\{[\sin(\pi x)]^{2(p-1)}\right\} * \left\{[\sin(\pi x)]^{2(p-1)}\right\} \approx \int_0^1 [\sin(\pi x)]^{4(p-1)}dx = \frac{1\times 3\times 5\cdots(4p-5)}{2\times 4\times 6\cdots(4p-4)}. \tag{E.26}$$

We can treat k_p as a constant using Eq. (E.26) with little loss of accuracy in most cases. For $p = 3$ and $\overline{N} = 10$, the convolution drops only about 0.4 dB by the time the linear term exceeds the excess noise. The drop is about 1.3 dB for $p = 4$ and 2.5 dB for $p = 5$. The drop is less at higher $\overline{N}$, especially at lower values of p.

The values of k_p obtained by approximating the convolution as its zero-frequency value are shown under Eq. (2.4).

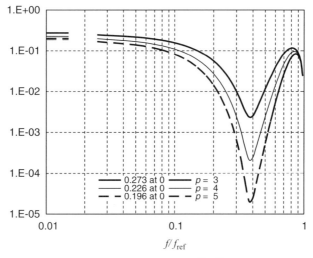

FIGURE E.1 Self-convolution of $[\sin(\pi x)]^{2(p-1)}$ over a range of $x = 1$.

E.3 APPROXIMATION IN EQ. (E.8)

Ignoring the terms beyond the third is valid as long as

$$\frac{2\pi|n\Delta t|}{T} = 2\pi|x\Delta t| \ll 1, \tag{E.27}$$

which implies

$$|x| \ll \frac{1}{2\pi|\Delta t|} \tag{E.28}$$

(note that we are using two-sided spectrums here, so x can be negative). Since $\Delta t = (n_c/F_{out})$ cycle, where n_c is the number of carries from the modulator, this can also be written as

$$|f_m| \ll \frac{F_{out}}{2\pi|n_c|} = \frac{NF_{ref}}{2\pi|n_c|}. \tag{E.29}$$

This restriction will usually be met up to the modulation frequency where the excess noise that we are computing is overpowered by the quantization noise from the linear term.

Equation (E.29) may not be met at $f_m = F_{ref}$, so Eq. (E.8) may not be useful in computing discrete reference spurs due to $\Sigma\Delta$ modulation (Appendix R).

APPENDIX F

REFERENCES TO *FS2*

Equations, figures, tables, or sections that are numbered "F.N" refer to those items numbered "N" in *FS2* (i.e., Egan [2000]: *Frequency Synthesis by Phase Lock*, 2nd edition).

Examples:

- Eq. (F.4.5) is Eq. (4.5) in *FS2*
- Fig. F.6.A.1 is Fig. 6.A.1 in *FS2*
- Table F.8.4 is Table 8.4 in *FS2*
- Section F.5.6 is Section 5.6 in *FS2*
- Chapter F.7 is Chapter 7 in *FS2*

Advanced Frequency Synthesis by Phase Lock, First Edition. William F. Egan.
© 2011 John Wiley & Sons, Inc. Published 2011 by John Wiley & Sons, Inc.

APPENDIX G

USING `Gsmpl`

G.1 OPEN-LOOP TRANSFER FUNCTION

G.1.1 Without Sampling

In the absence of sampling, the open-loop transfer function $G(\omega_m)$ is given by

$$G_0(\omega_m) = \frac{K}{\sec^{T-1}} h \frac{(1+j\omega_m/\omega_{z1})(1+j\omega_m/\omega_{z2})\cdots}{(j\omega_m)^T(1+j\omega_m/\omega_{p(T+1)})(1+j\omega_m/\omega_{p(T+2)})\cdots}, \qquad \text{(G.1)}$$

where T is the loop type and $h=1$. If there is sampling and a zero-order hold (as with a sample-and-hold phase detector),

$$h = \frac{1-e^{j\omega_m/f_s}}{j\omega_m/f_s} = e^{-j\pi f_m/f_s}\frac{\sin(\pi f_m/f_s)}{\pi f_m/f_s} \qquad \text{(G.2)}$$

is a first-order approximation for the effect of the sample-and-hold on the fundamental modulation frequency.

For a type-1 loop (one zero-frequency pole), the value of K in Eq. (G.1) is the same as K/N in *FS2* and K in *Phase-Lock Basics* (*PLB*), where $N=1$. For a type-2 loop, K in Eq. (G.1) equals ω_n^2 sec, the square of the natural frequency (except for the "sec" unit factor).

Advanced Frequency Synthesis by Phase Lock, First Edition. William F. Egan.
© 2011 John Wiley & Sons, Inc. Published 2011 by John Wiley & Sons, Inc.

G.1.2 Using Gsmpl

Gsmpl allows us to specify the parameters in Eq. (G.1) and to obtain $G(\omega_m)$ for a specified range of frequencies. A command line window opens and we are invited to

Enter Command (? for help)

(the parenthetical expression is a constant reminder). If we do enter ?, we obtain a list of commands, which is expanded below.

d: display parameters gives a list of current parameter values such as is shown in Fig. G.1.

ra causes radian units to be used, whereas Hz causes Hz units (entries are case sensitive). If a pole is entered in rad/s, for example, and Hz is then entered, the pole's units will be changed from rad/s to Hz and the numerical value will change to maintain the same frequency.

p+, p-, z+, z-: add or delete a pole or zero. If p+ is entered, for example, the program asks for the pole value and if z- is entered, it asks for the number of zero to be deleted. Note that the pole and zero numbers are displayed in Fig. G.1. [The poles at 0 need not have the lowest indices, as might be inferred from Eq. (G.1).]

h: add or delete (zero-order) hold circuit, toggles the existence of a hold circuit.

K: gain constant, as shown in Eq. (G.1). This would include the $1/N$ factor in a synthesizer loop.

fs is the sampling frequency, which equals the reference frequency F_{ref} (since we are approximating it as constant).

nf is the number of frequencies at which the response is computed, *beginning* at fb and *ending* at fe. We are asked to enter frequencies in the units that are in use, regardless of the use of "f" in these symbols.

G.1.3 Sampling Effects

When the effects of sampling are accounted for, the transfer function becomes

$$G(f_m) = G_0(f_m) + [G_0(f_m + f_s) + G_0(f_m - f_s)] + [G_0(f_m + 2f_s) + G_0(f_m - 2f_s)] + \cdots,$$

$$(G.3)$$

```
K = 707/sec
Sample Frequency is 7070 rad/sec
nP = 2
    pole # 1 is 0 rad/sec
    pole # 2 is 1414 rad/sec
nZ = 0
no hold circuit
20 frequencies from 100 to 10000 rad/sec
    in increments of 521.053 rad/sec or multiples of 1.27427
    using 0 harmonics.
```

FIGURE G.1 Displayed parameters.

where the number of harmonic pairs (2 are shown in Eq. (G.3)) is given by nh. Often results do not change with the use of more than a few pairs but, typically, hundreds of harmonic pairs can be included without noticeable delay.

ru will run and display the results at equally spaced frequencies. rg will space them geometrically (logarithmically).

G.2 CLOSED-LOOP RESPONSES

The closed-loop modulation response at the modulation frequency f_m is given by

$$H(f_m) = \frac{f_{\text{div}}(f_m)}{f_{\text{ref}}(f_m)} = \frac{G_0(f_m)}{1 + G(f_m)}, \tag{G.4}$$

where f_{div} is the output of the frequency divider, $G_0(f_m)$ is given by Eq. (G.1), and $G(f_m)$ is given by Eq. (G.3), although there are also outputs at other aliased frequencies $(nf_s \pm f_m)$. See Eq. (F.7.28).

The gain to the VCO output is $NH(f_m)$, where N is the divider ratio. The loop error response at modulation frequency f_m is

$$E(f_m) = 1 - H(f_m). \tag{G.5}$$

G.3 SAVING RESULTS

Parameters can be saved with the command s, which elicits a prompt for a file name. This saves the file to that name, allowing the parameters to be recovered by the command re, followed by the file name. Saving parameters makes it easier to restart the problem without having to enter all the parameters individually.

The Windows XP and Windows7 operating systems save files in, and read files from, the same directory that the program is in, but Macintosh System 10.5 saves to, and reads from, the user's main directory. For this reason, with Macintosh, it may be desirable to create a folder, named MyFolder, for example, at the user's top level and to enter the file's name as MyFolder/FileName. Then the files will be saved in and read from that folder. Of course, in any case, we can write a more complete path name if we take the time.

G.4 VERSION NUMBER

The program name may include a version number and file extension, for example, Gsmp12.exe.

G.5 EXAMPLE SESSION

Figure G.2 shows part of a Gsmp1 session.

```
    Enter Command (? for help)z+
  Enter VALUE, in Hz, of zero to ADD.
82.243
addZero of value 82.243

    Enter Command (? for help)d
K = 650000/sec
Sample Frequency is 1000 Hz
nP = 3
   pole # 1 is 0 Hz
   pole # 2 is 0 Hz
   pole # 3 is 482 Hz
nZ = 1
   zero # 1 is 82.243 Hz
no hold circuit
11 frequencies from 20 to 2000 Hz
   in increments of 198 Hz or multiples of 1.58489
   using 0 harmonics.

    Enter Command (? for help)rg
```

Closed Loop (dB)*			------------Open Loop Transfer Function------------				
Error	Gain/N	Freq.(Hz)	magnitude	(in dB)	degrees	real	imaginary
-32.33	0.20	2.0000e+01	4.232e+01	32.53	-168.71	-4.1506e+01	-8.2876e+00
-24.39	0.49	3.1698e+01	1.752e+01	24.87	-162.69	-1.6730e+01	-5.2155e+00
-16.54	1.08	5.0238e+01	7.603e+00	17.62	-154.53	-6.8645e+00	-3.2695e+00
-8.95	2.09	7.9621e+01	3.566e+00	11.04	-145.31	-2.9325e+00	-2.0299e+00
-2.16	3.10	1.2619e+02	1.832e+00	5.26	-137.76	-1.3563e+00	-1.2313e+00
2.30	2.30	2.0000e+02	9.997e-01	-0.00	-134.89	-7.0549e-01	-7.0824e-01
3.11	-2.16	3.1698e+02	5.452e-01	-5.27	-137.88	-4.0434e-01	-3.6567e-01
2.09	-8.98	5.0238e+02	2.796e-01	-11.07	-145.48	-2.3035e-01	-1.5841e-01
1.08	-16.58	7.9621e+02	1.309e-01	-17.66	-154.71	-1.1835e-01	-5.5925e-02
0.48	-24.44	1.2619e+03	5.673e-02	-24.92	-162.82	-5.4197e-02	-1.6752e-02
0.20	-32.39	2.0000e+03	2.347e-02	-32.59	-168.80	-2.3025e-02	-4.5571e-03

```
    *Modulation responses at input Freq.  Aliased outputs appear at other frequencies.

    Enter Command (? for help)nh
  Enter number of harmonics.
200
numH = 200

    Enter Command (? for help)rg
```

Closed Loop (dB)*			------------Open Loop Transfer Function------------				
Error	Gain/N	Freq.(Hz)	magnitude	(in dB)	degrees	real	imaginary
-35.28	0.15	2.0000e+01	4.260e+01	32.59	-168.79	-4.1789e+01	-8.2839e+00
-27.43	0.35	3.1698e+01	1.779e+01	25.01	-162.97	-1.7014e+01	-5.2098e+00
-19.76	0.77	5.0238e+01	7.857e+00	17.91	-155.48	-7.1488e+00	-3.2603e+00
-12.46	1.51	7.9621e+01	3.797e+00	11.59	-147.95	-3.2183e+00	-2.0153e+00
-5.69	2.53	1.2619e+02	2.041e+00	6.20	-143.73	-1.6458e+00	-1.2076e+00
0.42	3.49	2.0000e+02	1.207e+00	1.63	-146.35	-1.0045e+00	-6.6858e-01
4.61	2.70	3.1698e+02	7.862e-01	-2.09	-158.06	-7.2926e-01	-2.9379e-01
4.56	-2.32	5.0238e+02	6.350e-01	-3.94	-180.29	-6.3502e-01	3.2099e-03
0.86	-13.93	7.9621e+02	1.183e+00	1.46	-213.36	-9.8836e-01	6.5065e-01
0.16	-18.39	1.2619e+03	9.251e-01	-0.68	-152.01	-8.1688e-01	-4.3410e-01
-0.00	-251.54	2.0000e+03	8.859e+10	218.95	-180.00	-8.8592e+10	-3.8515e+05

```
    *Modulation responses at input Freq.  Aliased outputs appear at other frequencies.
```

FIGURE G.2 Part of a `Gsmpl` session. `Gain/N` equals *H*. Earlier versions do not compute closed-loop parameters.

G.6 GENERATING ANALYSIS PLOTS

Bode and Nyquist plots can be generated, by copying the program output to a spreadsheet. [Similar plots can also be obtained from the z-transform representation of the loop (see Section F.7.3) using $z = e^{sT_s}$, where $T_s = 1/f_s = 2\pi/\Omega_s$ and $s = \sigma + j\omega$.]

The first step is to copy the tables of results, or part of them, to the spreadsheet, giving a result as seen in Fig. G.3.

This was done on a Macintosh Computer by copying the two tables from the Gsmpl output, pasting each into a cell in the spreadsheet, and using Excel's *Data/Text to Columns* feature to place each data item in a cell. Alternatively, data can be copied one column at a time using option-drag to highlight each column before copying.

Magnitude (in decibels) and degrees were then plotted against frequency (Hz) to produce the Bode plots shown in Fig. G.4.

Data showing sampling effects are not plotted beyond about 500 Hz. With sampling, gain curves are symmetrical about half of the sampling rate and phase curves are symmetrical about the point defined by that frequency and $-180°$. However, they can look distorted due to the logarithmic (increasing) data spacing if they are plotted beyond half the sampling rate.

Error	Gain/N	No Sampling Freq.(Hz)	magnitude	(in dB)	degrees	real	imaginary
-32.33	0.2	2.00E+01	4.23E+01	32.53	-168.71	-4.15E+01	-8.29E+00
-24.39	0.49	3.17E+01	1.75E+01	24.87	-162.68	-1.67E+01	-5.22E+00
-16.54	1.08	5.02E+01	7.60E+00	17.62	-154.53	-6.86E+00	-3.27E+00
-8.95	2.09	7.96E+01	3.57E+00	11.04	-145.31	-2.93E+00	-2.03E+00
-2.16	3.1	1.26E+02	1.83E+00	5.26	-137.76	-1.36E+00	-1.23E+00
2.3	2.3	2.00E+02	1.00E+00	0	-134.89	-7.05E-01	-7.08E-01
3.11	-2.16	3.17E+02	5.45E-01	-5.27	-137.88	-4.04E-01	-3.66E-01
2.09	-8.98	5.02E+02	2.80E-01	-11.07	-145.48	-2.30E-01	-1.58E-01
1.08	-16.58	7.96E+02	1.31E-01	-17.66	-154.71	-1.18E-01	-5.59E-02
0.48	-24.44	1.26E+03	5.67E-02	-24.92	-162.82	-5.42E-02	-1.68E-02
0.2	-32.39	2.00E+03	2.35E-02	-32.59	-168.8	-2.30E-02	-4.56E-03
Error	Gain/N	With 1 kHz Sampling Freq.(Hz)	magnitude	(in dB)	degrees	real	imaginary
-35.28	0.15	2.00E+01	4.26E+01	32.59	-168.79	-4.18E+01	-8.28E+00
-27.43	0.35	3.17E+01	1.78E+01	25.01	-162.97	-1.70E+01	-5.21E+00
-19.76	0.77	5.02E+01	7.86E+00	17.91	-155.48	-7.15E+00	-3.26E+00
-12.46	1.51	7.96E+01	3.80E+00	11.59	-147.95	-3.22E+00	-2.02E+00
-5.69	2.53	1.26E+02	2.04E+00	6.2	-143.73	-1.65E+00	-1.21E+00
0.42	3.49	2.00E+02	1.21E+00	1.63	-146.35	-1.00E+00	-6.69E-01
4.61	2.7	3.17E+02	7.86E-01	-2.09	-158.06	-7.29E-01	-2.94E-01
4.56	-2.32	5.02E+02	6.35E-01	-3.94	-180.29	-6.35E-01	3.21E-03
0.86	-13.93	7.96E+02	1.18E+00	1.46	-213.36	-9.88E-01	6.51E-01
0.16	-18.39	1.26E+03	9.25E-01	-0.68	-152.01	-8.17E-01	-4.34E-01
0	-214.6	2.00E+03	1.26E+09	182.01	-180	-1.26E+09	-4.59E+04

FIGURE G.3 Data imported to a spreadsheet.

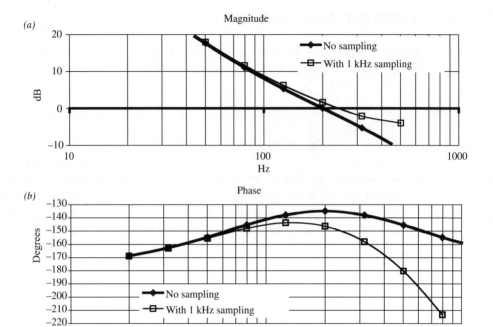

FIGURE G.4   Bode gain and phase plots.

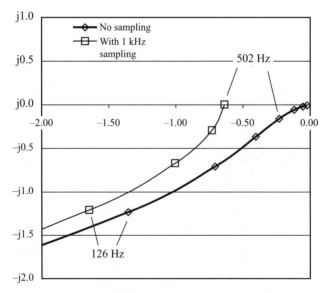

FIGURE G.5   Nyquist plots.

By a similar process, we can plot "imaginary" versus "real" to obtain a Nyquist plot (Fig. G.5).

G.7 VERIFICATION OF GARDNER'S STABILITY LIMITS

Gardner [2005, p. 275] gives a formula, Eq. (12.12), for gain at the stability limit for a type-2 third-order loop with no hold as a function of the frequencies of the loop filter's zero and pole and the sampling frequency. It can be written as

$$\frac{K_g}{\omega_z} = \frac{(\Omega_s/\omega_z)^2}{\pi^2(1+(\Omega_s/\pi\omega_z)(1-(\omega_z/\omega_p))\tanh(\pi\omega_p/\Omega_s))}, \tag{G.6}$$

where K_g is a variable K defined by Gardner and Ω_s, ω_p, and ω_z are sampling, pole, and zero frequencies, respectively. In terms of the K defined in Eq. (G.1), this is

$$K = \frac{\Omega_s^2}{\pi^2(1+(\Omega_s/\pi\omega_z)(1-(\omega_z/\omega_p))\tanh(\pi\omega_p/\Omega_s))}. \tag{G.7}$$

For one set of variables that satisfies this equation, Gsmpl data produce the portion of the Nyquist plot in Fig. G.6.

The plot goes right through $(-1, 0)$, showing exactly zero gain and phase margins, in agreement with Gardner's prediction. The instability occurs at half of the sampling frequency. The plot is shown with lower magnification in Fig. G.7, along with a plot at twice this unstable sampling rate and one with no sampling (nh $= 0$).

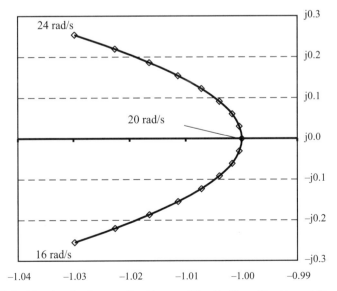

FIGURE G.6 Nyquist plot for $\omega_z = 2$ rad/s, $\omega_p = 20$ rad/s, $\Omega_s = 40$ rad/s, and $K = 25.91792$. Five thousand sideband pairs were used for accuracy.

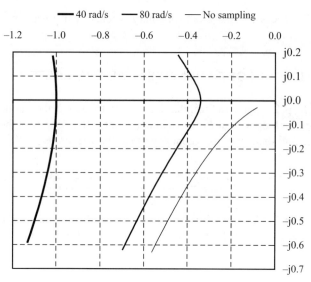

FIGURE G.7   Expanded Nyquist plot with additional curves for higher sampling rate and for no sampling.

At twice the unstable sampling frequency (middle curve), the gain margin goes to $1/0.34 = 2.94$ (9.4 dB). This again occurs at half of the (new) sampling frequency. A family of such curves could be produced at various sampling frequencies. When no sidebands are included (right curve), the effect of sampling is removed. The plot is still symmetrical about the x-axis, but only half of it is seen here because negative frequencies have not been used. With sampling, the plot repeats itself at increments of Ω_s, so negative frequencies are not required to see results on both sides of the x-axis. The region between Ω_s and $\Omega_s/2$ is the same as would be produced by negative frequencies.

G.8 THE NYQUIST PLOT

Here, we will briefly describe the theory of the Nyquist plot, with and without sampling, with particular attention to the type-2 third-order loop just discussed. First, we will list some facts about the Nyquist plot [*PLB*, Section 5.2.2]:

- $G(s)$ is assumed to be stable, so it has no RHP (right-half plane) poles;
- The poles of $G(s)$ are also the poles of $1 + G(s)$, so this also does not have RHP poles.
- The poles of $H(s)$ are zeros of $1 + G(s)$, so they occur at $G(s) = -1$.
- The number of zeros of $1 + G(s)$ encircled by the locus of s in the s plane equals the number of times that the corresponding locus in the $[1 + G(s)]$ plane encircles the origin, which is also the number of times that $(-1, 0)$ is encircled in the $G(s)$ plane.

Then we will look at the Nyquist plot for a type-2 third-order loop.

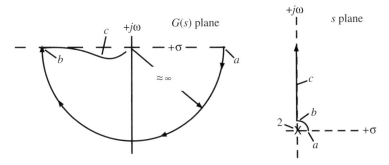

FIGURE G.8 Nyquist plot without sampling. The scale is necessarily distorted.

G.8.1 Without Sampling

The right-hand side of Fig. G.8 shows the s plane with the two poles of a type-2 loop at the origin. The zero and the other pole are off the visible area to the left. To encircle the RHP zeros of $[1 + G(s)]$ in the s plane, we begin at a very small value at a on the $+\sigma$ axis and circle counter-clockwise to b on the $+j\omega$ axis. This is done to avoid the poles at the origin; we avoid poles on the $j\omega$ axis so that they will not affect the plot in the $[1 + G(s)]$ plane. This very small $-90°$ arc in the s plane produces a very large $+180°$ arc, from a to b, in the $G(s)$ plane due to the $1/s^2$ term in $G(s)$.

As the locus in the s plane moves up the $+j\omega$ axis, the radius (magnitude) of $G(s)$ decreases and the zero in $G(s)$ produces positive phase shift that causes the locus to drop below the $-\sigma$ axis, avoiding the critical point at $(-1, 0)$, which is somewhere on that axis. The locus in $G(s)$ decays toward zero as the locus in s moves up to infinity.

The locus of s then circles at infinite radius to the $-j\omega$ axis and back up to point a, avoiding the origin by completing the half circle to a. The corresponding locus in the $G(s)$ plane completes a mirror image, mirrored in the σ axis. If the locus of $G(s)$ at c would have gone above $(-1, 0)$, the half of the locus that is shown would have encircled $(-1, 0)$, implying a RHP zero, and the mirror image would have done so again, indicating the second zero in a complex pair. That would have corresponded to complex poles of $H(s)$ in the RHP, an indication of instability. However, that cannot occur with the assumed loop.

G.8.2 With Sampling

The right-hand side of Fig. G.9 shows the s plane again, but this time we see the two poles at the origin repeated at $(0, j\Omega_s)$. Because of sampling, all the poles and zeros are now repeated every Ω_s in the vertical direction. The locus shown for $G(s)$ is completed as the s locus moves from a to c and then the mirror image is completed as s moves to d. So, if the $(-1, 0)$ point should be to the right of c in the $G(s)$ plane, that point would be encircled as s moves from a to d. (Note the symmetry about c in both planes.) The process would repeat continuously as s moved up the $j\omega$ axis, indicating an RHP zero of $G(s)$ for each Ω_s increment, consistent with the repeated poles in $H(s)$ with

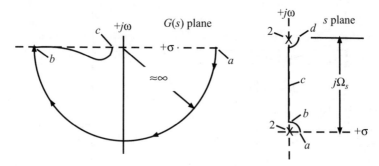

FIGURE G.9   Nyquist plot with sampling. The scale is necessarily distorted.

sampling. Conversely, if $(-1, 0)$ should be to the left of c in the $G(s)$ plane, it would not be encircled, indicating a stable system.

The process by which the Nyquist plot reveals RHP poles in $H(s)$ with sampling may be obvious from the description above, but we may wish to complete the formal process of encircling the RHP zeros of $G(s)$. Perhaps we should begin at a very large real value of s and move to a from there. Then $G(s)$ would begin at zero and move to a.[G1] Similarly, we could move s from d to a very large $+\sigma$ at constant $j\omega = j\Omega_s$ and then to $j\omega = 0$ at that large radius, completing a closed path encircling the RHP between $j\omega = 0$ and $j\omega = j\Omega_s$. The corresponding locus in the $G(s)$ plane would start at $(0, 0)$, move to a, follow the path shown to c, and then complete a symmetrical path. The locus in the $G(s)$ plane will still encompass $(-1, 0)$ if c is to the left of that point, indicating a zero in the segment of the RHP between $j\omega = 0$ and $j\omega = j\Omega_s$. Or, we could follow a similar closed path after moving far up the $j\omega$ axis in the s plane, encompassing one RHP zero for each segment.

APPENDIX H

SAMPLE-AND-HOLD CIRCUIT

Here we will look at the transient performance of the sample-and-hold circuits shown in Figs. 2.5 and 2.23 and of a modification of the former and we will consider the requirement for a capacitor between the charge pump and sampler.

We will find that the circuit in Fig. 2.5 has problems when the synthesizer steps down in frequency, especially for large steps. For this reason, we have added an inhibit to the circuit that is shown in the literature [Liu and Li, 2005; Cassia et al., 2003] to enable the sampling function to be overridden during frequency switching (acquisition). We will see that the ideal S&H circuit in Fig. 2.23 does not exhibit these problems and that the circuit of Fig. 2.5 can be used as a practical implementation of the S&H with a slight modification to its control signals.

H.1 TRANSIENT PERFORMANCE

We will compare the performance of these circuits in the synthesizer described in Fig. F.10.12, where simulation results from the MATLAB script SynCP.m are shown. The parameters given there are as follows:

$$K_p = 2\,\text{V/cycle}; \ K_{LF} = 10^5; \ K_v = 10^6\,\text{Hz/V}; \ F_{ref} = 10\,\text{kHz}.$$

Advanced Frequency Synthesis by Phase Lock, First Edition. William F. Egan.
© 2011 John Wiley & Sons, Inc. Published 2011 by John Wiley & Sons, Inc.

The loop filter has a pole at 1 kHz and a zero at 100 Hz. There is another pole at 0.001 Hz, but we will vary that very low-frequency pole to suit our circuits while maintaining the same gain above 1 Hz.

H.1.1   No Sampling

The response without a sampler (with the switch in Fig. 2.23 bypassing the sampler) is shown in Fig. H.1. The frequency, shown stepping from 5 to 10 MHz in the upper trace of Fig. H.1, is the same as can be seen in Fig. F.10.12b. A step back to 5 MHz is also shown in Fig. H.1. The middle waveform is the voltage following integration of the charge pump pulses.

H.1.2   Ideal Sampler

We now reposition the switch in Fig. 2.23 to insert an ideal S&H circuit. The PD block contains the integrator of the loop filter. We can see the resulting response in Fig. H.2. We note that the response is little affected by the sampling except for the introduction of some fine structure.

H.1.3   Hold with Integrator

We now change the PFD-and-integrator circuit to incorporate current sources and a capacitor, rather than a mathematical $1/s$ function, and the rest of the sampler circuit shown in Fig. 2.5, whose function is illustrated by the waveforms in Fig. 2.6, but without the cancellation pulse. The response to the previously simulated frequency change is shown in Fig. H.3. Note that the frequency does not drop when the frequency divider is changed from 1000 to 500 at 11 ms. We can see the reason in Fig. H.4.

The halving of the divider ratio causes the divider output (lower trace) to begin to arrive each 50 μs rather than the normal 100 μs ($1/F_{ref}$). As a result, a divider output pulse, Div, causes a D (pump-down) pulse to begin before Ref goes to 0. In Fig. 2.6, imagine the second Div pulse moving to the left so that it starts before Ref drops. This would activate D, which would not end until Ref, which terminates it, went to 1. Thus, D would last for the whole time Ref was down. As a result, the logical function $\overline{Ref \cdot D}$ (bottom of Fig. 2.6 and middle trace in Fig. H.4) is never true and the hold switch never closes. Since D is set at 1 by Div and this occurs every half period of Ref and since D is reset only when Ref becomes 1, there is no time when $\overline{Ref \cdot D}$ equals 1. (The second trace of Fig. H.4 shows D as essentially always on after 11 ms, but this is due to the timing established when the loop was locked at the higher value of N and is not essential to the problem.)

The voltage on the integrating capacitor continues to be charged negatively by the charge pump (fourth trace in Fig. H.4), but this is never transferred to the integrator in the hold circuit. It just continues to charge negatively until it is limited due to circuit restrictions. In any case, we are not seeing the response we want from the frequency change command.

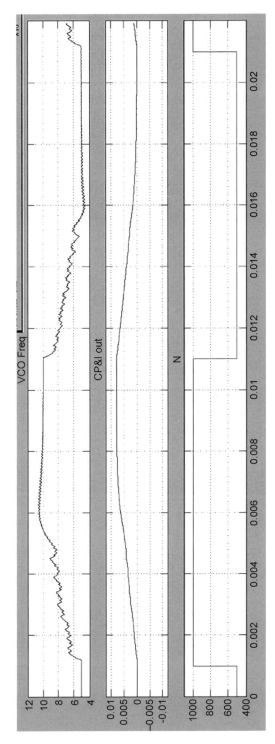

FIGURE H.1 Response without sampling, steps between 5 and 10 MHz, 2 ms/div. *Top to bottom:* frequency, voltage after the integrator, *N* switching between 500 and 1000.

197

198

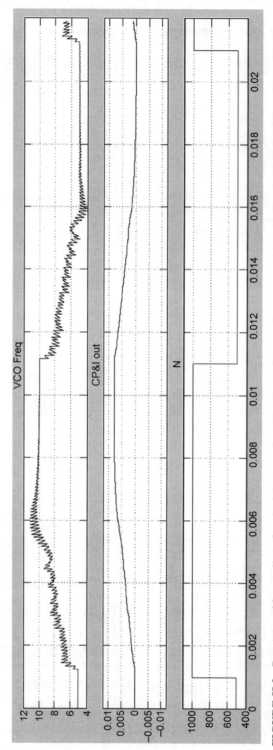

FIGURE H.2 Response with sampling after the integrator, 2 ms/div. *Top to bottom*: frequency, voltage after the **S&H** circuit (Fig. 2.23), *N* switching between 500 and 1000.

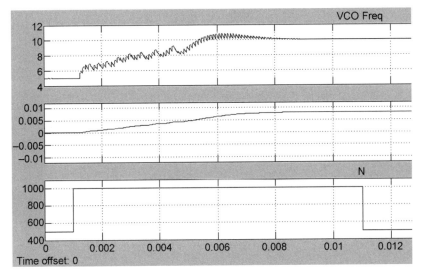

FIGURE H.3   Response with circuit of Fig. 2.5, step from 5 to 10 MHz and back, 2 ms/div. *Top to bottom*: frequency, held output, *N*.

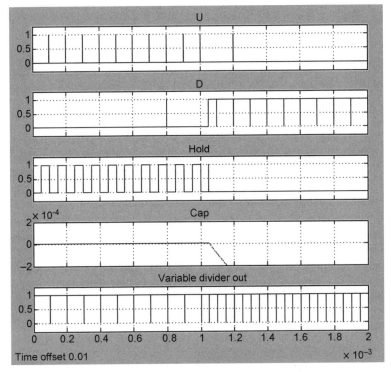

FIGURE H.4   Details during down step in Fig. H.3 between 10 and 12 ms. *Top to bottom*: U, D, connect-to-hold command, voltage on integrating capacitor, divider output.

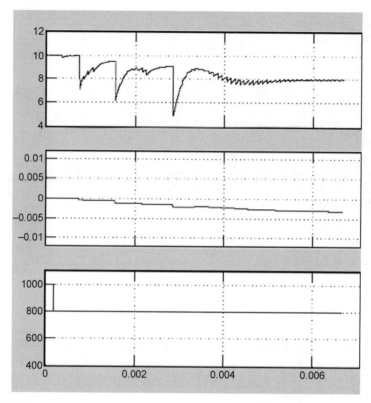

FIGURE H.5   Response to step from 10 to 8 MHz, 2 ms/div. *Top to bottom*: frequency, held output, *N*.

The hold circuit performs as intended as long as the synthesized frequency is not changed significantly. We can see what happens with a smaller step in Fig. H.5. There are periods when the *D* pulse keeps the hold from occurring, as before, but there are periods when the intended charge transfer can occur. The sudden transfer of charge to the hold circuit, at the beginning of the latter periods, causes a huge frequency transient. One can be seen in Fig. H.5 extending all the way to 5 MHz, which also would probably cause clipping or saturation or some other nonlinearity in a synthesizer that was not designed for such a frequency range.

Details of the time between 2 and 4 ms in Fig. H.5, the period containing the last (third) transient shown, are shown in Fig. H.6. Note how the integrator capacitor voltage is discharged regularly, when the circuit is performing as desired, on the right side of the figure.

H.1.4   Modified Hold with Integrator

The problem observed here results from the hold inhibiting that occurs during the pump-down signal. In normal operation, this arrangement allows the capacitor

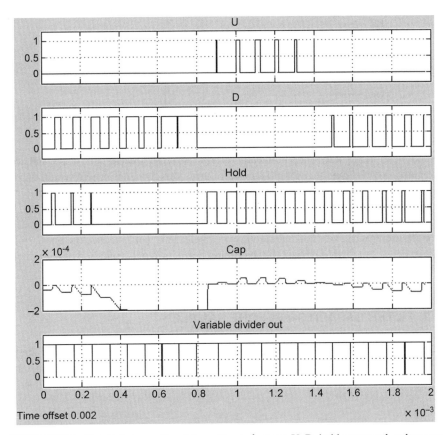

FIGURE H.6   Details between 2 and 4 ms. *Top to bottom*: U, D, hold command, voltage on integrating capacitor, divider output.

to charge for the longest time possible after the beginning of $\overline{\text{Ref}}$, but it can also prevent it from charging in response to a pump-down event (D) that begins before $\overline{\text{Ref}}$. Ideally, sampling occurs instantly at the beginning of $\overline{\text{Ref}}$, as with the ideal S&H, but practically it will require a finite time. If sampling is going to occur effectively at the desired time, the charging of the hold capacitor should be completed rapidly; so it should be acceptable to terminate the charge transfer from integrator capacitor to hold capacitor sooner than the beginning of the normal D pulse. Therefore, a modification is introduced to Fig. 2.5, as shown in Figs. H.7 and H.8, where the hold switch is closed from the beginning of $\overline{\text{Ref}}$ for a duration of $1/(4F_{\text{ref}})$. This signal can be generated by a monostable multivibrator or by using logic delays or, most likely, through logical combination of states in a frequency divider that produces F_{ref}, as suggested in the figure.

The response with the modified circuit, as shown in Fig. H.9, is essentially the same as with the ideal S&H in Fig. H.2. (This does not imply that circuit design considerations could not cause the S&H to respond more slowly than the simulated circuit.) Details of the down step are shown in Fig. H.10. The main change here is that

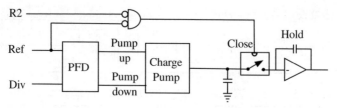

FIGURE H.7   Modified sample-and-hold circuit.

the connection is made to the hold each reference period independent of the state of the pump down signal. If D should still be 1 during this time, a transient would occur at the hold output, but the intention is that this would not occur when spectral purity is critical, during steady-state operation.

The hold switch could be closed closer to the reference time, for example, between a quarter reference cycle and half a reference cycle. This would reduce the phase shift due to delay, but further restrict the maximum correction pulse width. Neither effect, which we have discussed elsewhere, would be significant in many designs.

H.2 FILTER CAPACITOR BEFORE SAMPLER

When the output of a PFD and charge pump is sampled, there must be a capacitor between the charge pump and the sampling switch. The capacitor voltage is constant except during the usually brief period when the charge pump is charging it, so a

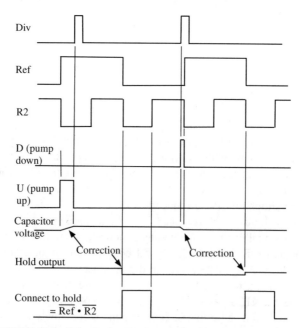

FIGURE H.8   Waveforms in the modified sample-and-hold circuit.

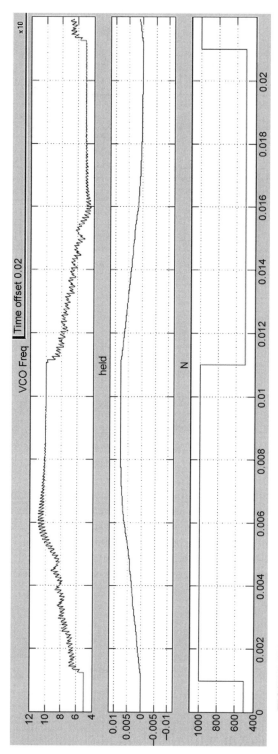

FIGURE H.9 Response with modified circuit of Fig. H.7, step from 5 to 10 MHz and back. *Top to bottom:* frequency, held output, *N*.

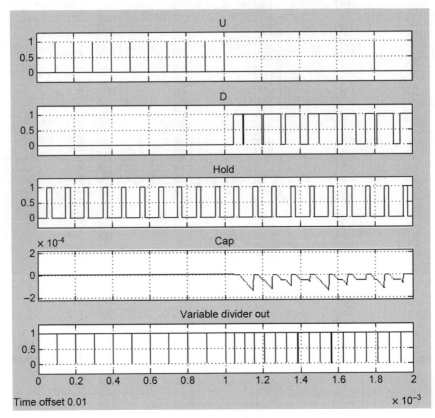

Time offset 0.01

FIGURE H.10   Details of the down step in Fig. H.9 from 10 to 12 ms. *Top to bottom*: U, D, hold command, voltage on integrating capacitor, divider output.

sample-and-hold output will look like the capacitor voltage except for a time delay. If the charge pump were to drive a resistor, the voltage pulse across the resistor typically would have disappeared by the time the sample would be taken.

It is common to place a resistor in series with the capacitor of the loop filter (Fig. H.11; we are showing tangential approximations of the frequency response) to provide a lead response, which is essential for loop stability. However, this too would

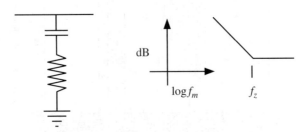

FIGURE H.11   Typical integrator-and-lead circuit.

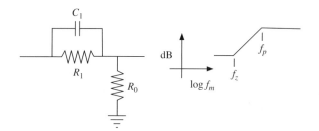

FIGURE H.12   Separate lead circuit.

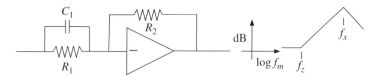

FIGURE H.13   Lead circuit with op-amp.

be ineffective between the charge pump and the sampler, for the same reason. The capacitor voltage would reflect the same change $\Delta V = I_{cp}\Delta T/C$ in response to the charge pump current I_{cp} that it would without the resistor, and the voltage across the resistor would be gone by the time the sample occurred. Therefore, the required lead circuit must be located after the sampler.

A lead circuit consisting of a resistor in series with the capacitor has the advantage of producing a transfer impedance that becomes constant at frequencies above f_z. A lead circuit that is separate must produce an ever-increasing response with frequency to counter the ever-decreasing response of the capacitor. Practically, its gain must level off at some frequency, thus creating a pole. The passive lead circuit in Fig. H.12 produces such a response. The gain begins to rise at $\omega_z = 1/(R_1 C_1)$ and flattens out when it reaches unity, the highest gain a passive circuit can have, at $\omega_p = 1/[(R_1 \| R_0)C_1]$.

We may also attempt to establish a lead with a shunt RC network at the input to an op-amp circuit (Fig. H.13). The gain will climb linearly with frequency until it again reaches the highest gain it can have, which is here the open-loop gain of the amplifier. The amplifier gain will typically be falling linearly with frequency, due to internal

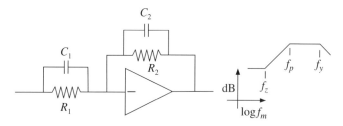

FIGURE H.14   Lead-lag circuit with op-amp.

compensation,[H1] and two gains will meet at f_x. At this frequency, the feedback circuit for the op-amp will be a voltage divider consisting approximately of C_1 and R_2 and will have a phase shift of nearly $-90°$. Combined with the transfer function of the amplifier (k/s), the open-loop transfer function for the op-amp circuit will be approximately $1/s^2$ (i.e., unity gain and $-180°$ phase), the condition for oscillation. If, on the other hand, the feedback circuit is shunt RC (Fig. H.14), a pole will be established, beyond which the total circuit gain will become flat. This flat gain will become equal to the falling amplifier's gain at a higher frequency f_y, one that ought to be well beyond the synthesizer's loop bandwidth. At this frequency, the open-loop response of the op-amp circuit will have 90° phase margin, because well beyond f_p, the op-amp's feedback circuit will be approximately a capacitive divider $[C_2/(C_1 + C_2)]$ with zero phase shift.

APPENDIX L

LOOP RESPONSE

L.1 PRIMARY LOOP

Unity gain is at 58 kHz and there is a filter zero at 16.8 kHz and two filter poles at about 310 kHz plus a steep, low-pass, elliptic filter with 1 MHz bandwidth.

L.1.1 Open-Loop Transfer Function

The forward gain is

$$G_F(s) = \frac{K_p}{s} F_{\text{LF}} K_v$$

$$= \frac{1/\text{cycle}}{s} \left[\frac{1.38 \times 10^{12}(s + 1.057 \times 10^5)}{s(s + 1.94 \times 10^6)(s + 1.95 \times 10^6)} EF(5, 1, 120, 1\,\text{MHz}) \right] 10^7\,\text{Hz}$$

$$= \frac{1.38 \times 10^{19}(s + 1.057 \times 10^5)}{s^2(s + 1.94 \times 10^6)(s + 1.95 \times 10^6)} EF(5, 1, 120, 1\,\text{MHz}), \tag{L.1}$$

where $EF(5, 1, 120, 1\,\text{MHz})$ is the transfer function of a 1 MHz bandwidth fifth-order elliptic filter with 1 dB in-band ripple and 120 dB out-of-band attenuation. Figure L.1

Advanced Frequency Synthesis by Phase Lock, First Edition. William F. Egan.
© 2011 John Wiley & Sons, Inc. Published 2011 by John Wiley & Sons, Inc.

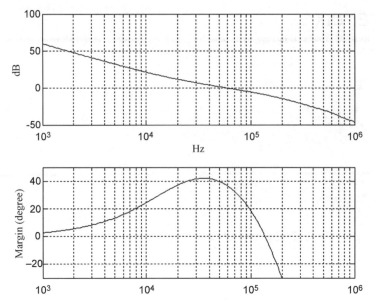

FIGURE L.1 Open-loop gain and potential phase margin (i.e., 180° + phase). From MATLAB script `opent58.m`.[L1]

shows the open-loop gain and phase of G_F/N, where we will be using $N = 10$ for the divide number.

We encounter a delay equal to half a cycle at 10 MHz in some of our circuits. This decreases the phase margin 1.0° at 58 kHz ($180° \times 58 \times 10^3/10^7$).

This loop has only 37° phase margin and 9 dB gain margin. While this has not been a problem for our experiments, it will cause overshoot in the time and frequency responses and could cause it to be subject to instability due to other parameter variations if it were used in a practical application. Without the elliptic filter, the phase margin would be a much more generous 53° and it would be 45° if the filter cutoff frequency were just doubled to 2 MHz. The existing, lower, corner frequency and the resulting narrower output spectrum facilitate observation of the spectrum by keeping images and aliased spectrums away.

L.1.2   Error Transfer Function

The error transfer function (Fig. 1.1*b*) is

$$\frac{E}{R} = \frac{f_e}{f_{\text{ref}}} = \frac{1}{1 + G_F/N} \tag{L.2}$$

and the gain (Fig. L.2) is its absolute value.

The effect of a delay that equals half a cycle at 10 MHz is small (see Fig. L.3).

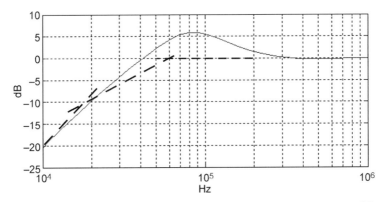

FIGURE L.2   Closed-loop error response. From MATLAB script `errort58.m`.[L2] Straight line approximations would be closer if the phase margin were greater.

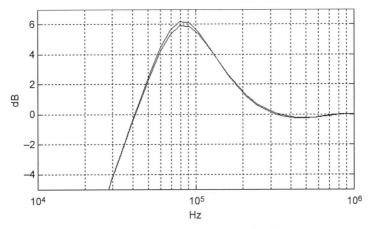

FIGURE L.3   Effect of a delay that equals half a cycle at 10 MHz on error response. Upper trace is with delay from MATLAB script `errort58delay.m`.[L3]

L.1.3 Forward Transfer Function

The transfer function (Fig. 1.1*b*) is

$$\frac{C}{R} = \frac{f_{\text{out}}}{f_{\text{ref}}} = \frac{G_F}{1 + G_F/N} = NH \tag{L.3}$$

and the gain divided by N is $|H|$, where

$$H = \frac{G_F/N}{1 + G_F/N}. \tag{L.4}$$

This is shown in Fig. L.4a and b. The straight line approximation to the log plot would be closer if the phase margin were greater. The linear plot is useful in analyzing spectral sideband levels.

The script ct58lin(1) produces $H(f)$ with a half cycle delay at f_{ref}, whereas ct58lin(0) produces $H(f)$ without the delay. Two scripts can be used to get the forward gain at a single frequency. Typing ct58at(f) gives the value of $|H(f_m = f)|$ with no delay, whereas typing ct58dat(f) gives the value of $|H(f_m = f)|$ including the delay.

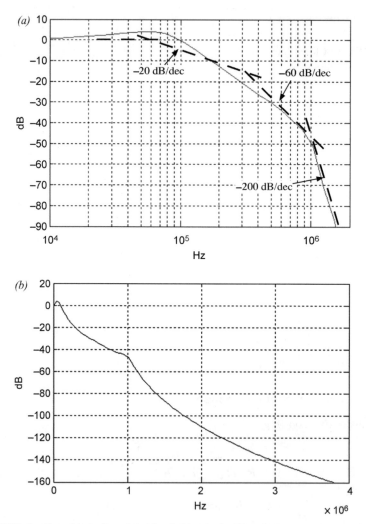

FIGURE L.4 Closed-loop forward gain divided by N, $|H|$. Log-frequency plot in (a) from MATLAB script closedt58.m. Linear frequency plot in (b) from MATLAB script ct58lin.m[L4] [type ct58lin(0)].

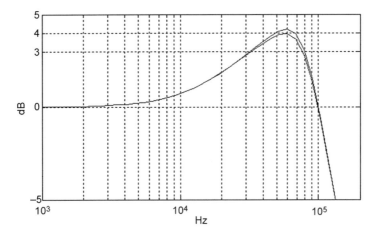

FIGURE L.5 Effect of a delay that equals half a cycle at 10 MHz. Lower trace is without delay from MATLAB script `closedt58.m`. Upper trace is with delay from MATLAB script `closedt58delay.m`.

The effect of a half cycle delay at 10 MHz on the closed-loop forward gain is small. See the log-frequency plots in Fig. L.5.

L.1.4 Output PPSD Shape

The theoretical output PPSDs with MASH modulators of various orders are shown in Fig. L.6. They can be compared with the various simulated responses, where the frequencies in Fig. L.6 equal the offsets from spectral center in the simulated

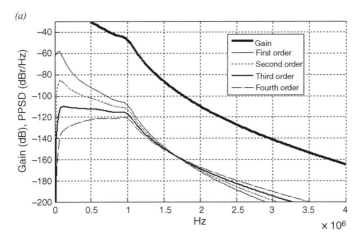

FIGURE L.6 Theoretical output PPSD. Linear frequency scale at (*a*); log scale at (*b*). Parameter is order of MASH modulator. Plots assume flat noise spectrum for quantization noise source. For the first order, since source is flat, output has same shape as gain. Compare with Figs. 2.30–2.32.

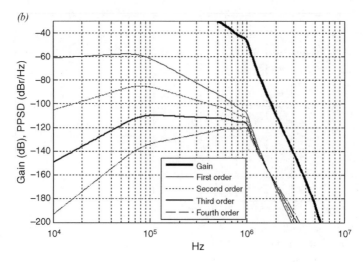

(b)

FIGURE L.6   (*Continued*)

responses. These were generated by the MATLAB scripts `sinlin58.m` and `sinlog58.m`, respectively. Type `sinlog58(n)`; where *n* is the order of the MASH modulator, to generate log-response curves like those in Fig. L.6*a*. A negative value for *n* produces the plot shown with five orders and the gain $|H(f_m)|$. Type `sinlin58(n,f2)` to generate a linear plot ending at $f_m = f2$, as in Fig. L.6*b*. Read the scripts for more details.

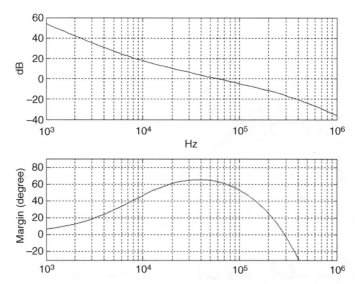

FIGURE L.7   Open-loop response of loop with greater phase margin from MATLAB script `opent58over.m`.[L5]

L.2 DAMPED LOOP

If the unity gain bandwidth were maintained at 58 kHz but the poles and zeros were moved twice as far from that frequency, we would obtain the following forward transfer function:

$$G_F(s)\frac{K_p}{s}F_{\mathrm{LF}}K_v = \frac{1/\mathrm{cycle}}{s}\left[\frac{5.65\times 10^{12}(s+0.528\times 10^5)}{s(s+3.88\times 10^6)(s+3.90\times 10^6)}EF(5,1,120,2\,\mathrm{MHz})\right]10^7\,\mathrm{Hz}$$

$$=\frac{5.65\times 10^{19}(s+0.528\times 10^5)}{s^2(s+3.88\times 10^6)(s+3.90\times 10^6)}EF(5,1,120,2\,\mathrm{MHz}).\qquad(\mathrm{L.5})$$

With $N=10$, this produces 63° phase margin and 16 dB gain margin, as can be seen in Fig. L.7.

Closed-loop responses now better match their straight line approximations, as can be seen in Figs. L.8 and L.9.

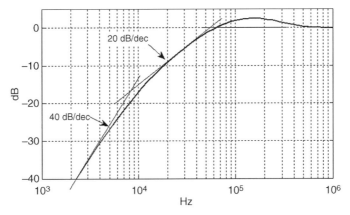

FIGURE L.8   Closed-loop error response with increased phase margin from MATLAB script errort58over.m.

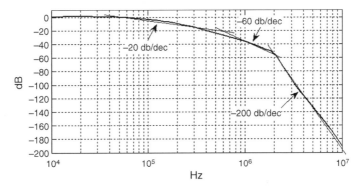

FIGURE L.9   Closed-loop forward response with increased phase margin from MALAB script closedt58over.m.

APPENDIX M

MASH PPSD

The PPSD due to a MASH modulator of order p is given by Miller and Conley [1991][M1] as

$$S_\varphi = \frac{(2\pi \text{ rad})^2}{6f_{\text{ref}}} \left[2\sin\left(\pi \frac{f_m}{f_{\text{ref}}} \right) \right]^{2(p-1)}, \qquad (\text{M.1})(2.3)$$

where the resulting output PPSD is

$$S_{\varphi,\text{out}} = S_\varphi |H(f_m)|^2. \qquad (2.2)$$

That is, Eq. (M.1) gives the output PPSD except for the modification at higher f_m due to the loop roll-off. Equation (F.8.74) is corrected to this form in the revised footnote 8. In the region of greatest concern, where $f_m \ll f_{\text{ref}}$, this reduces to Eq. (F.8.75), which is essentially the same as Miller and Conley's Eq. (15),

$$S_\varphi(f_m) \approx \frac{6.58}{f_{\text{ref}}} \left(\frac{2\pi f_m}{f_{\text{ref}}} \right)^{2(p-1)} \qquad (\text{M.2})$$

or, in decibel units,

Advanced Frequency Synthesis by Phase Lock, First Edition. William F. Egan.
© 2011 John Wiley & Sons, Inc. Published 2011 by John Wiley & Sons, Inc.

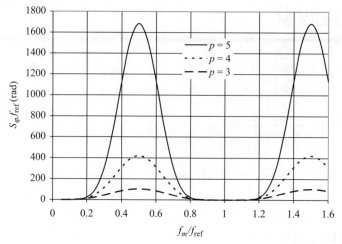

FIGURE M.1   MASH-11... quantization noise, linear plot, normalized. This is multiplied by $H(f_m/f_{ref})$ to give output PPSD due to MASH quantization.

$$S_\varphi(f_m)|_{dB} \approx [16p - 7.8 - 20\log_{10}(f_{ref}/Hz) + 20(p-1)\log_{10}(f_m/f_{ref})]dBr/Hz.$$

$$(M.3)$$

A normalized version of Eq. (M.1) is plotted in Figs. M.1 and M.2.
It is shown in Appendix Q that

$$S_\varphi = \frac{(2\pi\ rad)^2}{[2\sin(\pi(f_m/f_{ref}))]^2} S_{\delta N},$$

$$(Q.1)$$

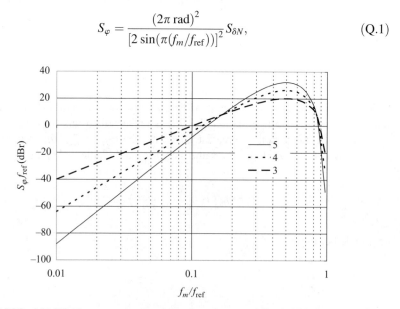

FIGURE M.2   MASH-11... quantization noise, log plot, normalized. This is multiplied by $H(f_m/f_{ref})$ to give output PPSD due to MASH quantization.

where $S_{\delta N}$ is the PSD of variations in the divide ratio, so the PPSD of Eq. (M.1) would result from a PSD at the output of the MASH modulator of

$$S_{\delta N} = \frac{1}{6f_{\text{ref}}} \left[2 \sin\left(\pi \frac{f_m}{f_{\text{ref}}} \right) \right]^{2p}. \tag{M.4}$$

The shape of $S_{\delta N}$ is proportional to the curves shown in Figs. M.1 and M.2, but with p higher by 1. In this appendix, we will show that Eq. (M.4) is true. This, combined with Eq. (Q.1), verifies Eq. (M.1).

Note that this appendix derives the shape of the MASH PSD from the modulator, whereas Appendix Q shows how this MASH PSD translates into a PPSD at the synthesizer output; so the two together show how Eq. (M.1) is produced.

M.1 MASH MODULATOR: FIRST STAGE

Figure M.3 shows the first stage of a MASH modulator. This diagram represents a real accumulator with capacity N_c. The block labeled "Accumulator" is a theoretical accumulator with infinite capacity. At each clock instant, it adds N_{in} to its previous contents. The output from the second summing junction is 0 until N_a reaches N_c because q equals $-N_a$ until then. When N_a becomes equal to or exceeds N_c, the value of q changes such that a value of N_c appears at c'. This causes a carry value of 1 to appear at the output. It also feeds back to the first summer so that, at the next clock, N_c will be subtracted from the accumulator contents. This represents the rollover of the real accumulator. The contents of the real accumulator is represented by $-q$, which tracks N_a but never exceeds N_c (Fig. M.4).

Figure M.5 is a more convenient variation of Fig. M.3 in which the numbers are divided by N_c. Thus, the numbers are all fractional, except for the carry, which is 1. To better understand how these diagrams relate to hardware implementations, see Appendix C.

Figure M.6 is a mathematical representation of Fig. M.5. The theoretical accumulator is represented by the transfer function

$$\frac{z^{-1}}{1-z^{-1}} = z^{-1} + z^{-2} + z^{-3} + \dots \tag{M.5}$$

This shows that the response to an input at one sample time is the value retained for all subsequent periods. This is accumulation or digital integration.

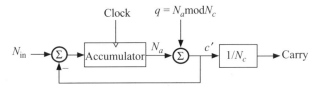

FIGURE M.3 First stage of a MASH modulator.

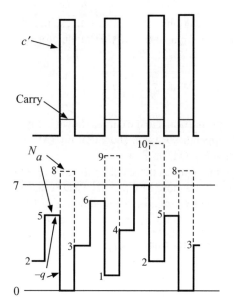

FIGURE M.4 Variables in Fig. M.3 for $N_c = 8$, $N_{in} = 3$.

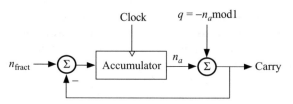

FIGURE M.5 First accumulator with number values referenced to its capacity.

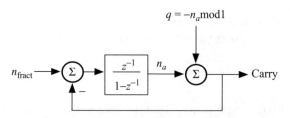

FIGURE M.6 Mathematical representation of Fig. M.5.

Applying the standard control system equation to Fig. M.6, we see that the closed-loop forward transfer function is

$$\frac{\text{carry}}{n_{\text{fract}}} = \frac{G}{1+G} = \frac{z^{-1}/(1-z^{-1})}{1+z^{-1}/(1-z^{-1})} = z^{-1} \tag{M.6}$$

and the response to q is

$$\frac{\text{carry}}{q} = \frac{1}{1+G} = \frac{1}{1+z^{-1}/(1-z^{-1})} = (1-z^{-1}).\qquad\text{(M.7)}$$

Therefore, the output is

$$\text{carry} = n_{\text{fract}}z^{-1} + q(1-z^{-1}).\qquad\text{(M.8)}$$

M.2 MASH MODULATOR: SECOND ORDER

When another accumulator is added for a second-order MASH, as in Fig. 2.10, the input to the second accumulator is the contents of the first accumulator, $-q_1$. Replacing n_{fract} in Eq. (M.8) by $-q_1$ (i.e., $-q$ from the first accumulator), we obtain the second carry output as

$$\text{carry}_2 = -q_1 z^{-1} + q_2(1-z^{-1}).\qquad\text{(M.9)}$$

As shown in Fig. 2.10, this is multiplied by $(1-z^{-1})$ and added to a delayed carry$_1$, the carry from the first accumulator, to produce

$$N = N_{\text{int}} + z^{-1}\text{carry}_1 + (1-z^{-1})\text{carry}_2\qquad\text{(M.10)}$$

$$= N_{\text{int}} + n_{\text{fract}}z^{-2} + q_1 z^{-1}(1-z^{-1}) + (1-z^{-1})[-q_1 z^{-1} + q_2(1-z^{-1})]\qquad\text{(M.11)}$$

$$= N_{\text{int}} + n_{\text{fract}}z^{-2} + q_2(1-z^{-1})^2.\qquad\text{(M.12)}$$

We have retained n_{fract} in the output, with an additional inconsequential delay, and we have replaced the quantization noise q_1 with quantization noise q_2. The difference is that, while q_1 just increases by n_{fract} each period, q_2, which is an accumulation of q_1, increases in a more complex fashion.

M.3 MASH MODULATOR: HIGHER ORDER

As we add more accumulators (Fig. 2.17), n_{fract} still appears in the output, but each added stage cancels the previous quantization noise and replaces it with a more complex sequence, just as occurred in Eq. (M.12). With p accumulators, it is easy to show (using the same process that we used above) that the total output is

$$N = N_{\text{int}} + n_{\text{fract}}z^{-p} + q_p(1-z^{-1})^p.\qquad\text{(M.13)}$$

When $p \geq 3$, it has been observed [Miller and Conley, 1991] that, except where n_{fract} contains only more significant bits (e.g., 0.25, 0.75), it becomes useful to represent q_p as a random number, equally distributed between 0 and 1. The variance of such a number is

$$\sigma_q^2 = \int_0^1 (x-0.5)^2 dx = \int_{-0.5}^{0.5} y^2 dy = \frac{1}{3} y^3 \Big|_{-0.5}^{0.5} = \frac{1}{12}. \tag{M.14}$$

This power is distributed over a (two-sided) frequency range of $\pm f_{\text{ref}}/2$, giving it a two-sided density$^{\text{M2}}$ of $S_{2q} = 1/(12 f_{\text{ref}})$. The sampling process replicates the spectrum at intervals of $F_s = F_{\text{ref}}$, causing the two-sided density everywhere to be $S_{2q} = 1/(12 f_{\text{ref}})$. This implies a one-sided density of

$$S_q = \frac{1}{6 f_{\text{ref}}}. \tag{M.15}$$

The representation of q as white noise when $p \geq 3$ is not just a reflection of the increased sequence length. The sequence length for $p = 3$ is only twice the length for $p = 1$ under the conditions of Miller and Conley's observation (Section X.2). The whiteness of q in the presence of a seed, a small initial value in the first accumulator, has been confirmed rigorously for $p \geq 3$, in contrast to $p < 3$, by Kozak and Kale [2004].

From Eq. (M.13), changes in N are

$$\delta N = q_p (1 - z^{-1})^p, \tag{M.16}$$

so the PSD of the changes is

$$S_{\delta N} = S_q \left| 1 - z^{-1} \right|^{2p} \tag{M.17}$$

$$= \frac{1}{6 f_{\text{ref}}} \left| 1 - e^{-j \omega_m T_s} \right|^{2p}, \tag{M.18}$$

where T_s is the sampling period, equal to $1/f_{\text{ref}}$. This can be further developed as

$$S_{\delta N} = \frac{1}{6 f_{\text{ref}}} \left| e^{-j \omega_m T_s/2} \left(e^{j \omega_m T_s/2} - e^{-j \omega_m T_s/2} \right) \right|^{2p} \tag{M.19}$$

$$= \frac{1}{6 f_{\text{ref}}} \left| 2 \left(\frac{e^{j \pi f_m / f_{\text{ref}}} - e^{-j \pi f_m / f_{\text{ref}}}}{2j} \right) \right|^{2p} \tag{M.20}$$

$$= \frac{1}{6f_{ref}} \left[2\sin\left(\pi \frac{f_m}{f_{ref}} \right) \right]^{2p},$$ (M.4)

which was to be shown.

M.4 VARIANCES

The corresponding variance is

$$\sigma_{\delta N}^2 = \int_0^{f_{ref}/2} S_{\delta N} df_m$$ (M.21)

$$= \int_0^{\pi/2} \frac{2^{2p}}{6\pi} \sin^{2p} x\, dx$$ (M.22)

$$= \frac{2^{2(p-1)}}{3} \frac{1 \cdot 3 \cdot 5 \dots (2p-1)}{2 \cdot 4 \cdot 6 \dots (2p)},$$ (M.23)

but it actually varies with n_{fract} when $p = 1$ (q_1 is not truly random).
   The sum (accumulation) of a sequence of values of δN is given by

$$\sum \delta N(z) = \frac{\delta N(z)}{1 - z^{-1}}$$ (M.24)

leading to

$$S_{\Sigma \delta N} = \frac{S_{\delta N}}{|1 - z^{-1}|^2}.$$ (M.25)

From Eq. (M.17), this can be written as a function of p as

$$S_{\Sigma \delta N}(p) = S_{\delta N}(p-1)$$ (M.26)

leading to

$$\sigma_{\Sigma \delta N}^2(p) = \sigma_{\delta N}^2(p-1),$$ (M.27)

where (p) again expresses a functional relationship (i.e., p is changed to $p-1$ in $\sigma_{\delta N}^2$ to get $\sigma_{\Sigma\delta N}^2$).

M.5 SOME PARAMETERS OF S_φ

From Eqs. (M.1) and (Q.1), we see that

$$S_\varphi(p,f_m) = (2\pi\,\text{rad})^2 S_{\delta N}(p-1,f_m),\qquad(\text{M.28})$$

so the integral of S_φ from $f_m=0$ to f_{ref} can be written as

$$\int_0^{f_{\text{ref}}} S_\varphi(p,f_m)df_m = (2\pi\,\text{rad})^2 \int_0^{f_{\text{ref}}} S_{\delta N}(p-1,f_m)df_m \qquad(\text{M.29})$$

$$= (2\pi\,\text{rad})^2 2\sigma_{\delta N}^2(p-1),\qquad(\text{M.30})$$

where the last equality is apparent from Eq. (M.21).

The noise bandwidth for S_φ', the part of S_φ between $f_m=0$ and f_{ref}, is

$$B_n(S_\varphi',p) = \frac{\int_0^{f_{\text{ref}}} S_\varphi'(f_m,p)df_m}{S_\varphi(f_m,p)_{\max}} = \frac{\int_0^{f_{\text{ref}}} S_{\delta N}'(f_m,p-1)df_m}{S_{\delta N}(f_m,p-1)_{\max}} \qquad(\text{M.31})$$

$$= \frac{2\sigma_{\delta N}^2(p-1)}{2^{2(p-1)}/(6f_{\text{ref}})} = 3f_{\text{ref}}2^{2(2-p)}\sigma_{\delta N}^2(p-1).\qquad(\text{M.32})$$

M.6 PREVIOUS DEVELOPMENT

The development in Section F.8.3 explicitly gives the equations for a third-order MASH and would be helpful for anyone who has difficulty with the transition from second order to any other order. One difference with that development is that the first-order accumulator is realized without a forward delay; the delay is in the feedback (Fig. C.6 rather than Fig. C.3). This simplified the development while showing the same fundamental result, but is less likely to correspond to a practical circuit.

M.7 SOME MASH MODULATOR CHARACTERISTICS

See Table M.1.

TABLE M.1 Some MASH Modulator Characteristics, $n_{fract} < 1$

Order	1	2	3	4
Maximum output	1	2	4	8
Minimum output	0	-1	-3	-7
$\sigma^2_{\delta N}$ output variance[a]	$\leq 0.25^b$	0.50^c	1.67	5.83
$\sigma^2_{\Sigma\delta N}$ integral variance[d]	0.083^c	0.167^c	0.50	1.67
|Integral|[e] and approximate probability excludes $n_{fract} = 0$				≥ 4, 0.0%
				≥ 3, 0.5%
		≥ 2, 0%	≥ 2, 0%	≥ 2, 6.5%
	≥ 1, 0%	≥ 1, 0.008%	≥ 1, 8.5%	≥ 1, 23%

14-bit accumulators with LSB set in the first accumulator. Some data differ at $n_{fract} = 0$, but modulator output could be inhibited there. 0% means 0 in 2^{15} clocks.
[a]From Eq. (M.23), excepting order 1, confirmed by `mashall2.m` or `mashall3.m` or `mashall4.m`.
[b]Depends on n_{fract}, maximum at 0.5.
[c]≈ 0 for $n_{fract} = 0$.
[d]From Eq. (M.27), confirmed by `mashall4.m`.
[e]Integral is running sum (accumulation) of outputs minus n_{fract}, the deviation in input cycle periods at the PD. Data from script `mashall3.m`. Probabilities are the same for positive and negative values.

M.8 CHARACTERISTICS OF MATLAB SCRIPTS MASHONE AND MASHALL_

See Table M.2.

TABLE M.2 Characteristics of MATLAB© scripts `mashone` and `mashall_`

		mashone	mashall	mashall2	mashall3	mashall4
PARAMETERS						
	Number	2	6	4	5	4
	Order	Input	Input	Input	Input	Input
	n_{fract}	Input	–	–	–	–
	fstart	–	Input	Input	Input	Input
	fpts	–	Input	Input	Input	Input
	fstop	–	Input	Input	Input	Input
	bits	14	Input	14	14	14
	m	15	Input	15	15	15
	nlim	–	–	–	Input	–
OUTPUTS						
	Type	Text	Plots	Plots	Plots	Plots
	Min, mean, max out	√	√	√	√	√
	out σ^2, theory	√	√	√	√	
	out σ^2, simulated	√	√	√	√	√
	σ^2, First quantizer in	√	–	–	–	–
	Frequency of ($\|$sum$\| \geq$ nlim)	–	–	–	√	–
	Sum σ^2, simulated	–	–	–	–	√

`sum` refers to the accumulated (digitally integrated) output. `fpts` values on n_{fract} are evenly spaced from `fstart` to `fstop`. `bits` is the number of bits in each accumulator. The number of clock cycles simulated at each n_{fract} is 2^m. The LSB in the first accumulator is initially set (1). Other bits are initially reset (0). Parameters are input in the order shown, top to bottom. For example, `mashall2(order, fstart, fpts, fstop)`.

APPENDIX N

SAMPLED NOISE

Consider a train of pulses of width T and frequency f_{ref} multiplying a noise waveform in the time domain. In the frequency domain, the noise power spectrum will be convolved with the Fourier power spectrum of the pulse train (Fig. N.1). See Section N.5 for clarification. The original noise spectrum at B will be replicated, each new replica being centered on one of the lines in the spectrum at A. We will consider three cases that differ in width of the noise spectrum compared to the width of the pulse spectrum and its repetition frequency f_{ref}.

N.1 CASE 1: $W_n \ll f_{ref}$

The noise power spectrum is narrow, for example, flicker noise that falls off with frequency. Unlike the flat spectrum shown at B in Fig. N.1, this spectrum is high in the center and falls off rapidly. Then, the other spectrums (C, D, etc.), resulting from convolution with the spectral lines at A, will have the same peaked shape and will not overlap significantly, and the noise seen near 0 will mainly just be the spectrum at B.

N.2 CASE 2: $1/T \gg W_n \gg f_{ref}$

This may represent phase noise that affects the switching of a logic signal only during the rise time of the signal. The pulses are very narrow, so the pulse spectrum is spread

Advanced Frequency Synthesis by Phase Lock, First Edition. William F. Egan.
© 2011 John Wiley & Sons, Inc. Published 2011 by John Wiley & Sons, Inc.

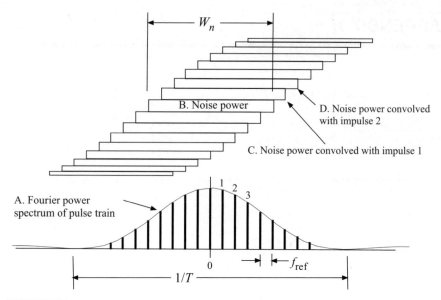

FIGURE N.1   The Fourier power spectrum of a pulse train is shown at A. At B is a noise power spectrum labeled "noise power." Above and below B are other noise power spectrums that result from the convolution of spectrum B with the impulses in spectrum A.

out and near the center looks like a flat impulse train. If the width of the noise spectrum is narrow (the frequencies are low) relative to the width of the pulse spectrum $(1/T)$, the total power near 0 will depend on how many spectral lines are in the width of the noise spectrum. Therefore, the total noise power density near 0 will be proportional to $1/f_{ref}$.

N.3 CASE 3: $W_n \gg 1/T \gg f_{ref}$

The pulses are relatively wide, so the pulse spectrum is narrow compared to the noise spectrum; each of the spectral lines will cause an aliased noise spectrum to add to the noise near 0 and the total noise there will depend on the total power in the pulse spectrum.

N.4 VARIANCE OF SAMPLED NOISE $(1/T \gg f_{ref})$

If a random process occurs at intervals of $1/f_{ref}$ and has a variance of σ^2, it can be modeled as sampled values from noise that has a two-sided density of $S_2'(f_m) = \sigma^2/(nf_{ref})$ over a bandwidth $\pm nf_{ref}/2$ for some value of n. Aliasing that occurs when the noise is sampled will produce a density $S_2(f_m) = n\sigma^2/nf_{ref} = \sigma^2/f_{ref}$, so we may as well consider the noise with variance σ^2 to be spread over $\pm f_{ref}/2$, since that will produce the same observable result [Egan, 2007, Appendix i17.M.1 or Egan,

1998, Appendix 18.M.1]. The corresponding one-sided density is $S(f_m) = 2\sigma^2/f_{ref}$ from 0 to $f_{ref}/2$. Of course, the noise could be in any other band of the same width. The observed results would be the same due to aliasing.

N.5 CONVOLUTION OF PSDs

When a noise waveform $n(t)$ is multiplied by a pulse train $p(t)$ to produce a sampled noise spectrum $s(t)$,

$$s(t) = n(t) \times p(t), \tag{N.1}$$

the Fourier transform of s is the convolution of the transforms of n and p,

$$S(f) = N(f)*P(f). \tag{N.2}$$

The convolution process equates $S(f)$ to the sum of all the products of components of $N(f')$ and $P(f')$ whose frequencies differ by f. (When we multiply two sinusoids, the result contains sum frequencies, but these result from differences between positive frequencies in one factor and negative frequencies in the other.) When one of the factors is a noise spectrum, the phases of the summed products are random. We cannot determine the resulting waveform without knowing these phases but we do know, because of the random phases, that the powers of the various products add. Therefore, we can write

$$|S(f)|^2 = |N(f)|^2*|P(f)|^2. \tag{N.3}$$

This expression states that the PSD of the resulting product is the convolution of the PSDs of the two factors. It results from adding the powers of the components generated by convolution rather than their values.

The process is relatively easy to follow in Fig. N.1, because convolution with a unit impulse $\delta(f-f_{imp})$ is particularly simple; the spectrum $N(f)$ is simply shifted, so zero frequency in $N(f)$ lies at the frequency of the impulse f_{imp},

$$N(f)*\delta(f-f_{imp}) = \int_{-\infty}^{\infty} N(f-x)\delta(x-f_{imp})dx = N(f-f_{imp}). \tag{N.4}$$

In Fig. N.1, $f_{imp} = mf_{ref}$ (m an integer); so the convolution produces, at f, the sum of $|N(f)|^2$ shifted by many multiples of f_{ref} (and multiplied by the squared value of each impulse). Again, since the phases of these values of $N(f)$ are random, we cannot find a value for $S(f)$, but we can find $|S(f)|^2$ and the PSD of $S(f)$.

We can think of the functions of f as discrete values, equal to the densities multiplied by a very narrow bandwidth, if that helps. The explanation is developed in these terms in Egan [2003, Section 5.1.3]. There, however, it is complicated somewhat by the fact that we discuss squaring, so the two factors are identical. This produces phase coherence in some of the products and leads to a factor of 2 in the resulting PSD.

N.6 REPRESENTING SQUARED PSDs

Briefly, if we represent the densities as sums of discrete terms, $\sum a_i$, the product of the two densities $\sum a_i$ and $\sum b_i$ contains pairs

$$c_j = a_j b_{j+k} + a_{j+k} b_j. \tag{N.5}$$

If the two densities are identical, however (squaring), $a_i = b_i$ and

$$c_j = a_j a_{j+k} + a_{j+k} a_j = 2 a_j a_{j+k}. \tag{N.6}$$

If the phases of a_i and b_i are random but their powers are the same for all i, c_j in Eq. (N.6) has twice the power of c_j in Eq. (N.5). That is why a factor of 2 appears when densities are squared, as in Eq. (E.19) (see Egan [2003; also Bracewell [1965], p. 337).

APPENDIX O

OSCILLATOR SPECTRUMS

Figure O.1 can also be considered to be Fig. F.3.38*f*. It shows additional oscillator spectrums from more recently available information. Data are shown in Table O.1, which is a continuation of Table F.3.1. Many of these curves are for IC oscillators, which tend to be relatively noisy. However, other oscillators that are used in ICs, such as ring oscillators and astable multivibrators, are even noisier and are included in Fig. F.3.38*e*.

Advanced Frequency Synthesis by Phase Lock, First Edition. William F. Egan.
© 2011 John Wiley & Sons, Inc. Published 2011 by John Wiley & Sons, Inc.

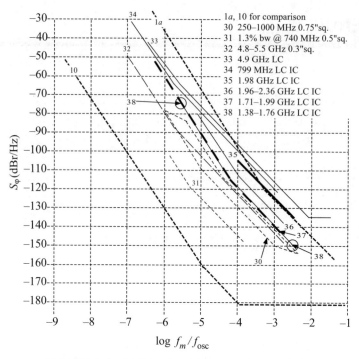

FIGURE O.1 Additional oscillator spectrums (F.3.38*f*).

TABLE O.1 (Table F.3.1 Continued)

Curve	Frequency	Description	Reference
1a[a]	240 MHz	15 dBm 240–352 MHz bipolar, maximum noise	FS2, p. 116; Egan [1972]
10[a]	500 MHz	SAW resonator lab prototype	FS2, p. 116; Montress and Parker [1990, Fig. 15, p. 530]
30	250 (top)–1000 MHz	Planer resonator, Synergy DCYR 25100-5 0.75 in. sq.	Rohde and Poddar [2006]
31	740 MHz	1.3% bandwidth, Z-Com ZRO 0743B2LF 0.5 in. sq.	Z-Communications, Inc. [2008]
32	4.8–5.47 GHz	Planer resonator, Synergy DRO473542-5 0.3 in. sq.	Rohde and Poddar [2008]
33	4.9 GHz	LC quadrature oscillator IC	van der Tang et al. [2002]
34	799 MHz	LC IC	Margarit et al. [1999]
35	1.98 GHz	LC IC	Zannoth et al. [1998]
36	1.96–2.36 GHz	LC IC	Andreani and Sjöland [2002]
37	1.71–1.99 GHz	LC IC	Tiebout [2001]
38	1.38–1.76 GHz	LC quadrature IC	Vancorenland and Steyaert [2002]

[a]These curves are included in all the plots of Fig. F.3.38 to provide a visual reference between plots. They are close to the upper and lower limits of oscillator spectrums excluding some of the very noisy, low-Q oscillators (e.g., ring oscillators).

APPENDIX P

PHASE DETECTORS

The sample-and-hold (S&H) phase detector (PD) [*FS2*, pp. 188–197] has been largely replaced by the phase frequency detector (PFD) driving a charge pump (CP) [*FS2*, pp. 197–211]. Historically, the PFD had the advantage of providing an acquisition aid [*FS2*, pp. 452–459] along with the PD in a single IC, whereas the S&H PD required additional logic circuitry [*FS2*, pp. 465–472] to perform that function. While this advantage seems to be lacking in a synthesizer IC, the CP PD is nevertheless almost universally used in ICs, for reasons that are not obvious. It may be easier to implement or it may have some performance advantage or, perhaps, design inertia plays a role.

The S&H PD provides a DC output in steady state and transients that can produce reference spurs are due to secondary effects, such as logic signals coupled capacitively or through ground currents. It provides stepped outputs in response to the shaped quantization noise in a $\Sigma\Delta$ synthesizer. This is easier to compensate than are the pulses from a PFD and it is easier to observe the phase with these levels, which is one reason that we have used it in some simulations.

The basic output from a PFD and CP is a pulse, with obvious implications for reference spurs. To avoid spurs, the PFD is normally used in a type-2 loop, so there will be zero phase error and, therefore, zero pulse width at steady state. Unfortunately, the PFD-and-CP tends to have a narrow nonlinear region just at zero phase [*FS2*, pp. 203–211, 233–242]. This can lead to unacceptable behavior, like that caused by a "dead zone" where the loop gain disappears. Moreover, the region is so narrow that it is hard to observe and the phase may wander in and out of the region as

Advanced Frequency Synthesis by Phase Lock, First Edition. William F. Egan.
© 2011 John Wiley & Sons, Inc. Published 2011 by John Wiley & Sons, Inc.

counteracting leakage currents change, with temperature for example. To avoid the region, it is now common to cause the PD to generate a pair of narrow canceling pulses at zero phase. However, these pulses carry noise and the noises do not cancel. Moreover, if the pulses are not matched exactly, they can cause reference spurs and can lead to errors in the designed quantization noise at the PD output and noise floors at the synthesizer output.

While the S&H PD avoids these problems, it may have some of its own due to finite response times [*FS2*, pp. 192, 193, 307–312], especially at high reference frequencies. We show (Section 2.3) how it can be important for the phase to change in synchronism with the reference frequency and how this can be accomplished with an unusual configuration of the S&H PD. We also show how a S&H circuit can be employed to resample the PD output in synchronism with the reference frequency to attain the desired synchronism. However, this S&H circuit can be subject to the same problems that potentially affect the S&H PD.

APPENDIX Q

QUANTIZATION PPSD

The phase power spectral density (PPSD) at the synthesizer output is

$$S_{\varphi,\text{out}} = S_\varphi |H(f_m)|^2. \tag{2.2}$$

In this appendix, we will show that

$$S_\varphi = \frac{(2\pi \, \text{rad})^2}{[2 \sin(\pi(f_m/f_{\text{ref}}))]^2} S_{\delta N}, \tag{Q.1}$$

where $S_{\delta N}$ is the PSD of the variation in divide number. The shape of S_φ with a MASH modulator, Eq. (2.3), is obtained by combining Eq. (Q.1) with $S_{\delta N}$, as derived in Appendix M and given there by Eq. (M.4).

The original form of Eq. (F.8.74) [revised as Eq. (2.3)] was obtained by employing the equivalence between a change in N and a frequency change introduced after the VCO,[Q1] as developed in Section F.2.5. However, this equivalence depends on steady state during the time preceding the change in N. While this applies to a single step in N, it is not accurate for modulated N. To obtain Eq. (Q.1) [and, therefore, also Eq. (2.3)] correctly, we must compute how changes in N produce changes in phase at the divider output. We will initially assume a constant

Advanced Frequency Synthesis by Phase Lock, First Edition. William F. Egan.
© 2011 John Wiley & Sons, Inc. Published 2011 by John Wiley & Sons, Inc.

synthesized frequency (which is our design goal). If the system were time independent, we could use superposition to add this result to the effects of changes in output
frequency (as with feedback or due to a change in $\overline{N}$). We will see that this is still a
good approximation in most cases.

Q.1 DEVELOPMENT OF Eq. (Q.1)

In the types of phase detectors used in synthesizers, u_1 (Fig. 1.1) is proportional to the
time difference between the PD trigger point on the reference waveform and the
trigger point on the divider output waveform. Since the period of the divider output is
N times the period T_{out} at the synthesizer output, a change δN from the mean divider
ratio $\overline{N}$ will cause the time of the divider output to deviate by

$$\delta t = T_{\text{out}} \delta N, \tag{Q.2}$$

and this will result in a PD output voltage change of

$$\delta u_1 = K_p \text{ cycle } \delta t / T_{\text{ref}} = K_p \text{ cycle } \delta N (T_{\text{out}} / T_{\text{ref}}) = K_p \text{ cycle } \delta N / \overline{N}, \tag{Q.3}$$

where T_{out} has been approximated as constant.

Similarly, the total PD voltage change after a series of k such incremental
deviations is

$$\Delta u_{1,k} \overset{\Delta}{=} \sum_{i=1}^{k} \delta u_{1,i} = \frac{K_p \text{ cycle}}{\overline{N}} \sum_{i=1}^{k} \delta N_i. \tag{Q.4}$$

The phase change at the divider output that would produce $\Delta u_{1,k}$ in our model is

$$\Delta \varphi_k = \frac{\Delta u_{1,k}}{K_p} = \frac{\text{cycle}}{\overline{N}} \sum_{i=1}^{k} \delta N_i. \tag{Q.5}$$

Therefore, it is appropriate to accumulate δN to obtain the value of $\Delta \varphi_k$. In the z-
domain,[Q2] we represent the accumulation of $\delta N(z)$ by dividing it by $(1 - z^{-1})$, so

$$\Delta \varphi(z) = \frac{\text{cycle}}{\overline{N}} \frac{\delta N(z)}{1 - z^{-1}}. \tag{Q.6}$$

The PPSD thereby introduced at the phase detector is

$$S_{\Delta \varphi, N} = \frac{\text{cycle}^2}{\overline{N}^2} \frac{S_{\delta N}}{|1 - z^{-1}|^2}. \tag{Q.7}$$

In the case of interest, $\Delta\varphi$ is an equivalent phase change injected at the divider output, equivalent to φ_{ref} in Fig. 1.1c (or $\Delta\varphi_N$ in Fig. F.6.A.1), and δN is a deviation of N from the mean due to quantization noise. Substituting $z = \exp(j\omega_m T_{ref})$, we obtain

$$S_{\Delta\varphi,N} = \frac{\text{cycle}^2}{\overline{N}^2} \frac{S_{\delta N}}{|1 - e^{-j\omega T_{ref}}|^2} \tag{Q.8}$$

$$= \left(\frac{2\pi\,\text{rad}}{\overline{N}}\right)^2 \frac{S_{\delta N}}{\left|e^{-jf_m/f_{ref}}\right|^2 \left|2\left((e^{jf_m/f_{ref}} - e^{-jf_m/f_{ref}})/2\right)\right|^2} \tag{Q.9}$$

$$= \left(\frac{2\pi\,\text{rad}}{\overline{N}}\right)^2 \frac{S_{\delta N}}{2\sin(\pi(f_m/f_{ref}))}. \tag{Q.10}$$

The output PPSD will then be

$$S_{\varphi,\text{out}} = S_{\Delta\varphi,N}\left|\overline{N}H(f_m)\right|^2 = \frac{(2\pi\,\text{rad})^2}{[2\sin(\pi(f_m/f_{ref}))]^2} S_{\delta N}|H(f_m)|^2, \tag{Q.11}$$

which is the quantization noise, Eq. (Q.1), multiplied by the loop response, as was to be shown. When $S_{\delta N}$ is given by Eq. (M.4), this becomes the output PPSD due to quantization given by Eqs. (2.2) and (2.3).

Q.2 SUPERPOSITION

Since the *system* is time dependent, we cannot assume superposition. The time of the kth transition in the divider output can be written as

$$T_k = \sum_{i=1}^{k} T_{\text{out},i} N_i = \sum_{i=1}^{k}[(\overline{T}_{\text{out}} + \delta T_{\text{out},i})(\overline{N} + \delta N_i)] \tag{Q.12}$$

$$= k\overline{T}_{\text{out}}\overline{N} + \sum_{i=1}^{k}[\delta T_{\text{out},i}\overline{N} + \overline{T}_{\text{out}}\delta N_i + \delta T_{\text{out},i}\delta N_i]. \tag{Q.13}$$

The change in T_k from nominal is then

$$\delta T_k \stackrel{\Delta}{=} T_k - k\overline{T}_{\text{out}}\overline{N} = \sum_{1}^{k} \overline{T}_{\text{out}}\delta N_i + \sum_{1}^{k} \delta T_{\text{out},i}\overline{N} + \sum_{1}^{k} \delta T_{\text{out},i}\delta N_i. \tag{Q.14}$$

The first term on the right represents the effect of the fractional-N modulation described in Eq. (Q.2), which results in an equivalent phase modulation at the loop input as described by Eqs. (Q.5)–(Q.8).

The second term is the effect of changes in the output period multiplied by the average value of N. This is assumed when we employ equations representing feedback, such as when the equivalent input modulation is multiplied by $\overline{N}|H(f_m)|^2$ in Eq. (Q.11). Typically, we would use $\overline{N}$ in developing $H(f_m)$.

The third term is a residual error, the sum of each change in output period multiplied by the corresponding deviation of N. The first two terms represent superposition of the effects of the divider modulation and the output period modulation. The third term is an error in the results of the superposition.

Since we are designing for small modulation sidebands, we expect that once the loop has settled at a given synthesized frequency,

$$\delta T_{\text{out},i} \ll \overline{T}_{\text{out}}, \qquad (Q.15)$$

so each member of the error summation would be small compared to the corresponding member in the first summation. That is,

$$\sum_{i=1}^{k}(\overline{T}_{\text{out}}\delta N_i + \delta T_{\text{out},i}\delta N_i) = \sum_{i=1}^{k}(\overline{T}_{\text{out}} + \delta T_{\text{out},i})\delta N_i \approx \sum_{i=1}^{k}(\overline{T}_{\text{out}}\delta N_i), \qquad (Q.16)$$

so we can drop the error term.

Similarly, if

$$\delta N_i \ll \overline{N}, \qquad (Q.17)$$

each member of the error summation would be small compared to the corresponding member in the second summation. That is,

$$\sum_{i=1}^{k}(\delta T_{\text{out},i}\overline{N} + \delta T_{\text{out},i}\delta N_i) = \sum_{i=1}^{k}\delta T_{\text{out},i}(\overline{N} + \delta N_i) \qquad (Q.18)$$

$$\approx \sum_{i=1}^{k}\delta T_{\text{out},i}\overline{N}, \qquad (Q.19)$$

another reason to drop the error term.

Q.3 NEW SYNTHESIZED FREQUENCY

Injection of an equivalent frequency step before the divider, as in Fig. 1.2b and Section F.2.5, is appropriate for analyzing the response to a simple change in N. (Approximations are discussed in Section F.2.5; see there footnote 3.) Use of the final value of N during the transient is exact in this case. When fractional modulation also exists, the final $\overline{N}$ is an approximation for that fixed final value, while δN_i is then a perturbation due to the fractional modulation.

Before the loop has settled, following a change in commanded frequency, (Q.15) may not hold true and we would depend on (Q.17) to allow the error term to be ignored [i.e., the approximation (Q.19)]. Even if (Q.17) is not very well met, the use of $H(f_m)$ based on the final $\overline{N}$ is probably a reasonable approximation.

Q.4 LOOP RESPONSE

When we account for sampling, $H(f_m)$ is given by Eq. (F.7.28) as^{Q3}

$$H(f_m) = \frac{G_0(f_m)}{1 + \sum_{n=-\infty}^{\infty} G_0(f_m + nf_{\text{ref}})}, \qquad (Q.20)$$

where G_0 is the open-loop transfer function without sampling. [This is the same as Eq. (G.4).] Within the loop bandwidth f_L (i.e., for $f_m < f_L$),

$$\sum_{n=-\infty}^{\infty} G_0(f_m + nf_{\text{ref}}) \approx G_0(f_m), \qquad (Q.21)$$

as long as the bandwidth is small compared to f_{ref} (i.e., $f_L \ll f_{\text{ref}}$), since the gain for nonzero values of n is small at those frequencies. The magnitude of the effect is discussed in Chapter F.7 and the exact transfer function can be computed, as it was there. As we go higher in f_m, the denominator becomes approximately unity until we come within the loop bandwidth of a multiple of f_{ref}. At that point, the denominator provides additional attenuation, becoming infinite at any multiple of f_{ref}, since $G_0(f_m + nf_{\text{ref}})$ becomes there $G_0(mf_{\text{ref}} + nf_{\text{ref}})$, which is infinite for $n = -m$.

Thus, the nulls in $S_{\Delta\varphi,\text{out}}$ (other than the one at $f_m = 0$) are accentuated by a response that further reduces the gain in their vicinity. However, since S_φ is already low in that region, and the summation of responses in the denominator of Eq. (Q.20) usually has little effect near the peaks of S_φ, the terms representing sampling in $H(f_m)$ can often be ignored. That is, the summation in the denominator of Eq. (Q.20) can often be replaced by $G_0(f_m)$, as suggested by Eq. (Q.21).

Nevertheless, there may be occasions when we want to employ Eq. (Q.20) to obtain an exact response, especially when f_L is pushed higher toward f_{ref}.

Q.5 VERIFICATION OF THE EFFECT OF SAMPLING ON THE LOOP

As a verification of Eq. (Q.20), we can observe the accentuation of the nulls in Fig. Q.1, which shows the frequency power spectral density (FPSD) of a synthesizer output in the vicinity of the null at $f_m = f_{\text{ref}}$ in Eq. (Q.11).

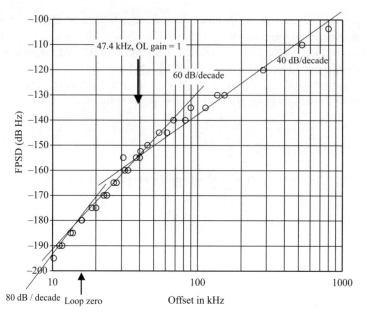

FIGURE Q.1   FPSD for a MASH-111 synthesizer versus offset from null at F_{ref}.

This FPSD is taken from the tuning voltage of the VCO in a simulated (via Simulink model) MASH-111 $\Sigma\Delta$ synthesizer. (The model contains no other noise sources that might have prevented the effect from being observed.) If we ignore sampling, the shape of this FPSD curve should be essentially the same as the shape of S_φ in Eq. (2.3), in spite of having been multiplied by $|H(f_m)|^2$ and the requirement to divide it by f_m^2 to change it to PPSD [see Eq. (F.3.36)]. Over the range of the graph, it would be reduced by another 0.8 dB if divided by f_m^2 to get PPSD, but removal of the effects of $|H(f_m)|^2$ would restore that 0.8 dB.$^{\text{Q4}}$

For small offsets from $F_{\text{ref}} = 10\,\text{MHz}$,

$$\delta_f = f_m - 10\,\text{MHz},$$

and $p = 3$, we can see that a slope of 40 dB per decade is produced; starting with Eq. (M.1),

$$S_\varphi = k_1 \left| 2\sin\left(\pi \frac{f_m}{f_{\text{ref}}}\right) \right|^{2(p-1)} = k_1 \left| 2\sin\left(\pi \frac{\delta_f}{10\,\text{MHz}}\right) \right|^{2(p-1)}$$

$$\approx k_1 \left(2\pi \frac{\delta_f}{10\,\text{MHz}}\right)^{2(p-1)} = k_2 \delta_f^4, \tag{Q.22}$$

$$S_\varphi|_{\text{dB}} = 10\,\text{dB}\,\log_{10} S_\varphi \approx k_3 + 40\,\text{dB}\,\log_{10} \delta_f. \tag{Q.23}$$

In the figure, we see that this 40 dB/decade slope at higher offsets δ_f, but as δ_f comes within the bandwidth of the loop, the term $G(f_m - nf_{\text{ref}} = \delta_f)$ in the denominator of Eq. (Q.20) becomes larger than unity and further increases the attenuation of $H(f_m)$. As is usual for well-behaved loops, $G(\delta_f) \sim 1/\delta_f$ in this region, so the slope increases by 20 dB/decade. When δ_f falls below the frequency of the loop zero, another increase of 20 dB/decade occurs due to the rising loop gain. Thus, the added dip in PPSD close to f_{ref} is evidence of Eq. (Q.20).

APPENDIX R

REFERENCE FREQUENCY SPURS

Here, we will show the theoretical levels of sidebands at $\pm F_{\mathrm{ref}}$ due to three PFD effects, and how they combine, and we will verify the theory with results from simulations.

First, we will perform an analysis of a train of narrow pulses, generated by the PFD, to be used in subsequent computations; we will assume no resampling until Section R.6. Then we will compute the first sideband amplitude due to leakage current (Section R.1). In Section R.2, we will compute the amplitude of sidebands due to an offset in two "canceling" pulses from a PFD. In Section R.3, we will compute the sideband amplitude due to $\Sigma\Delta$ modulation. In Sections R.4 and R.5, we will find the result of combining $\Sigma\Delta$ spurs with pulse offset spurs and with leakage current spurs, respectively, comparing theoretical and simulation results.

The amplitude of the fundamental component from the Fourier series of a train of rectangular pulses is

$$A_1 = \frac{2A_p}{\pi} \sin \pi D_p, \qquad (\mathrm{R.1})(\mathrm{F.5.55})$$

where A_p is the pulse amplitude and D_p is the pulse duty factor. Because the pulses that produce spurs are narrow compared to the repetition period, this is approximately

$$A_1 \approx 2A_p D_p. \qquad (\mathrm{R.2})$$

Advanced Frequency Synthesis by Phase Lock, First Edition. William F. Egan.
© 2011 John Wiley & Sons, Inc. Published 2011 by John Wiley & Sons, Inc.

The equivalent input peak phase deviation is

$$m_{1,\text{ref}} = \frac{A_1}{K_p} = \frac{2A_pD_p}{K_p} = \frac{2A_pD_p}{A_p/2\pi \text{ rad}} = 4\pi D_p \text{ rad.} \tag{R.3}$$

R.1 LEAKAGE CURRENT

Leakage current I_L at the input to the integrator following the CP will be canceled by a pulse from the CP, so the average current from the canceling pulse, I_pD_p, must equal I_L, and the amplitude of the fundamental component of the canceling pulse will therefore be [Eq. (R.2)]

$$I_1 \approx 2I_pD_p = 2I_L. \tag{R.4}$$

The amplitude of the equivalent input phase modulation would be

$$m_{1,\text{ref},L} = I_1/K_p \tag{R.5}$$

$$= 2I_L/K_p. \tag{R.6}$$

The first sideband at the output would be

$$SB_{1L} \approx \frac{m_{1,\text{out}}}{2} = \frac{m_{1,\text{ref}}NH(F_{\text{ref}})}{2} = \frac{I_L}{K_p}NH(F_{\text{ref}}), \tag{R.7}$$

where $m_{1,\text{out}}$ is the peak phase deviation of the fundamental component at the loop output.

R.2 PULSE OFFSET

Opposing narrow pulses of amplitude A_p that do not quite cancel due to a time offset of T_o (Figs. F.5.47 and F.5.48) lead to a Fourier component at F_{ref} with amplitude

$$A_1 \approx 4\pi A_pD_pT_o/T_{\text{ref}}. \tag{F.5.62}$$

Following the same procedure that produced Eq. (R.3), this leads to equivalent input phase modulation with a peak deviation

$$m_{1,\text{ref},o} = \frac{4\pi A_pD_pT_o/T_{\text{ref}}}{A_p/2\pi \text{ rad}} \tag{R.8}$$

$$= 8\pi^2 \text{ rad } D_pT_o/T_{\text{ref}}, \tag{R.9}$$

which would produce the first sideband of

$$SB_{1o} = (2\pi)^2 D_p \frac{T_o}{T_{ref}} NH(F_{ref}). \tag{R.10}$$

While both Eqs. (R.7) and (R.10) depend on F_{ref} because H changes with frequency, there is an additional dependence of the latter equation on F_{ref}^2, which can be written as

$$SB_{1o} = (2\pi)^2 \frac{T_p}{T_{ref}} \frac{T_o}{T_{ref}} NH(F_{ref}) = (2\pi)^2 T_p T_o \left(\frac{F_{ref}}{cycle}\right)^2 NH(F_{ref}). \tag{R.11}$$

Therefore, pulse offsets become more important at higher values of F_{ref} (unless T_p and T_o are proportional to T_{ref}).

Recall also that the signal producing SB_{1o} is in quadrature with the signal that produces SB_{1L}, so the two effects could not be adjusted to cancel each other (Section F.5.7.6).

R.3 $\Sigma\Delta$ MODULATION

The phase pulses generated by $\Sigma\Delta$ modulation are similar to the offset pulses considered previously. As illustrated in Fig. 2.6, positive CP pulses are offset to one side of the reference transition and negative pulses are offset to the other side (pump-up and pump-down pulses). Therefore, a Fourier series representing these pulses will have similarities to the series representing the offset pulses and Eq. (R.9) can be modified to give the resulting modulation index,

$$m_{1,ref,\Sigma\Delta} = 8\pi^2 \frac{T_p T_o}{T_{ref}^2} \text{ rad.} \tag{R.12}$$

In the $\Sigma\Delta$ case, at the ith reference period in a sequence,

$$T_p(i) = T_o(i) = n_d(i)/F_{out}, \tag{R.13}$$

where n_d is the number of pulse periods T_{out} by which the divider output is shifted. It equals the sequence of carry numbers n_c from the modulator after digital integration,

$$n_d = \frac{n_c}{1-z^{-1}}. \tag{R.14}$$

Since $n_d(i)$ varies over the sequence, we will use an average value

$$m_{1,ref,\Sigma\Delta} = \left(\frac{2\pi}{F_{out}T_{ref}}\right)^2 \overline{n_d^2} = \left(\frac{2\pi}{\overline{N}}\right)^2 \overline{n_d^2} = \left(\frac{2\pi}{\overline{N}}\right)^2 \sigma_{nd}^2. \tag{R.15}$$

We have also reduced Eq. (R.12) by a factor of 2, since unlike in Section R.2, only one pulse occurs per period. That is, it takes two periods to produce the pair of pulses from which Eq. (R.12) is derived. We have obtained σ_{nd}^2 from theory [Eq. (M.27)] and confirmed it by simulations (Section M.7). It is equal to 0.5 for MASH-111 and 1.67 for MASH-1111, giving

$$m_{1,\text{ref},\Sigma\Delta3} = 19.7 \, \text{rad}/N^2 \tag{R.16}$$

for MASH-111 and

$$m_{1,\text{ref},\Sigma\Delta4} = 65.8 \, \text{rad}/N^2 \tag{R.17}$$

for MASH-1111.

This implies equivalent input sidebands, for MASH-111, of

$$\text{SB}_{1,\text{ref},\Sigma\Delta3} = 19.9 \, \text{dBc} - 40 \, \text{dB} \, \log_{10} \overline{N} \tag{R.18}$$

and, for MASH-1111, of

$$\text{SB}_{1,\text{ref},\Sigma\Delta4} = 30.4 \, \text{dBc} - 40 \, \text{dB} \, \log_{10} \overline{N}. \tag{R.19}$$

These should also apply to other configurations that produce quantization noise given by Eq. (2.3).

R.4   EFFECT OF ΣΔ MODULATION ON PULSE OFFSET SPURS

Simulations were performed to observe the effect of ΣΔ modulation and offset pulses on the reference spurs. The system response was as described in Section L.1, except the elliptic filter was changed to a single pole to reduce the attenuation at F_{ref}.

In one test, delays of 1 and 3 ns were inserted in series with the turnoff signals to the two D flip-flops in the PFD logic in order to produce opposing 3 ns pulses offset from each other by 2 ns. (This is similar to Fig. F.5.35, where, however, the 1 ns delay is internal to both D flip-flops.) The response shown in Fig. R.1a was measured on the tuning line; it is calibrated in dB Hz, dB relative to 1 Hz^2/Hz of frequency power spectral density. Here, the noise bandwidth is 3749 Hz, so 10 dB $\log_{10}$ 3749 = 35.7 dB should be added to the apparent spur level to change from density to rms frequency deviation. When this is done, the deviation shown in Fig. R.1a corresponds to the A_1 of Eq. (R.1) (after accounting for the gain between the two points and the conversion from amplitude to two-sided spectral power).

The MASH-111 ΣΔ modulator was then turned on and the spur level increased by 14.1 dB (Fig. R.1b), representing an increase in deviation by five times. This new deviation is the sum of the original deviation caused by the offset spurs alone and the

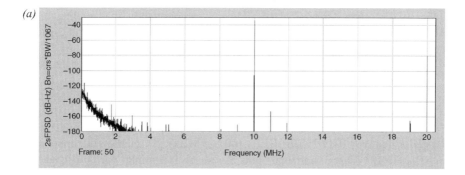

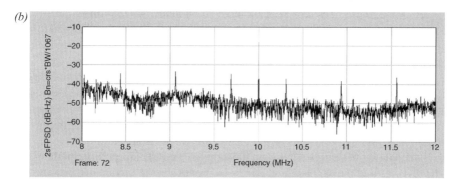

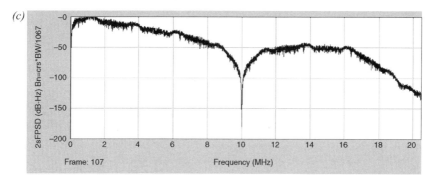

FIGURE R.1 Frequency deviation measured on the tuning voltage. Units are for density in dB Hz (Hz^2/Hz). A -33.2 dB Hz sideband due to pulse offset in the PD, shown in (a), increases 14.1 dB in (b) with MASH-111 modulation. In (c), resynchronization after the PD eliminates the spur. Add 10 dB log10 $Bn = 35.7$ dB for rms deviation relative to 1 Hz.

deviation that would be caused by the ΣΔ modulation alone, which we will now demonstrate. They add because the two sinusoids are almost in phase, each going through zero at the midpoint between pulses. This is at the reference transition for the ΣΔ modulation and is offset from that point by only 0.5 ns (half of the pulse overlap) for the offset pulses, which represents a phase difference of only 1.8°.

Figure R1*c* shows the effect of adding a sample-and-hold (sampling half a reference cycle after the reference transition) after the phase detector and capacitor output. The spur has disappeared (or been reduced by at least 160 dB).

In Fig. R.2, the lowest curve gives the theoretical level of the first sideband due to offset pulses with no $\Sigma\Delta$ modulation and the three flat line segments on the left are levels due to $\Sigma\Delta$ modulation without pulse offset. The other curves are sums of these two curves and the data points are measured values.

In more detail, the sideband level computed from Eq. (R.19) for the fourth-order modulator and $N = 10.0625$ (for $F_{out} = 100.625$ MHz) is shown by the upper short (from 0, 0 to 1, 1) dashed horizontal line. Below this is a similar line computed from Eq. (R.18) for a third-order modulator and the same N and then a solid line for $N = 20.0625$. These lines represent $\Sigma\Delta$ modulation alone, without offset pulses.

The dash-dot curve in the lower right is the computed value of the sideband for the offset pulses, from Eq. (R.8) (divided by 2 to change m to relative sideband level), representing sideband levels due to offset pulses without any $\Sigma\Delta$ modulation.

The other curves are the numerical sums of the theoretical levels for $\Sigma\Delta$ modulation alone and for offset pulses alone. All the curves that we have just described represent the theory given above. All the symbols are measured values (obtained by dividing the measured FPSD by the gain from the input). The correspondence between the measured values and the sums of the theoretical values is apparent.

If the delays are switched between flip-flops in the PFD, the polarity of the pulse pair reverses and the fundamental component opposes the fundamental due to the $\Sigma\Delta$ modulation. In this case, we observe the subtraction of the two deviations (not shown here).

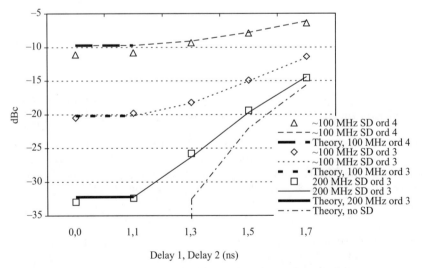

FIGURE R.2   Equivalent sideband level at input due to $\Sigma\Delta$ (SD) modulation and offset pulses. The *x*-axis is labeled with the delays in the two flip-flop resets. Curves are theoretical sums of spurs from the two effects. Symbols are measured values.

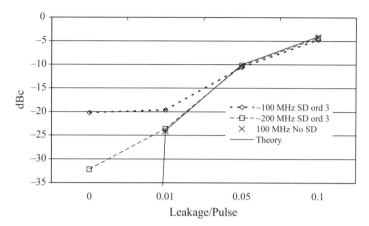

FIGURE R.3 Equivalent sideband level at input due to current leakage and $\Sigma\Delta$ (SD) modulation. Curves are theoretical power sums. Symbols are measured values.

R.5 EFFECT OF $\Sigma\Delta$ MODULATION ON LEAKAGE SPURS

Simulations were performed to observe the effect of simultaneous $\Sigma\Delta$ modulation and current leakage on the reference spurs under conditions as described previously. The results are shown in Fig. R.3. The lower solid curve represents the theoretical level of the equivalent input relative sideband level, due to leakage, from Eq. (R.5) (divided by 2). The other curves again represent the sum of this value with the theoretical level for $\Sigma\Delta$ modulation without leakage. This time, however, powers (mean square values) rather than deviations are added because the sinusoid due to leakage peaks at the center of the pulse that is created in response to the leakage, close to the reference transition. This sinusoid is approximately in quadrature with the sinusoid due to $\Sigma\Delta$ modulation. Again, we see the correspondence between theory and measured results.

R.6 EFFECTS OF RESAMPLING

Resampling eliminates all the effects that we have observed. The resampled level does not contain any of the pulses that caused the reference sidebands. Second-order effects due to transients in the sample-and-hold circuit will undoubtedly be present to some degree, but these are not subject to analysis without consideration of circuit details.

APPENDIX S

SPECTRUM ANALYSIS

Power spectrums of simulated waveforms are measured in spectrum analyzers using the fast Fourier transform (FFT) method of generating digital Fourier transforms (DFTs). An FFT is a DFT. Lyons [2004], especially pp. 120–122, is recommended as a reference.

S.1 SPECTRUMS

S.1.1 Periodicity

Since $\Sigma\Delta$ modulation is periodic, resulting waveforms are periodic. What we call quantization noise at the synthesizer output is actually a periodic waveform (assuming the synthesizer has settled). It can be characterized by a Fourier series consisting of harmonics of sinusoids at frequencies cycle/T_{sequ}, where T_{sequ} is the duration of the repeated sequence. If a Fourier series is obtained from a segment of a waveform of duration T_{segm}, it will represent a waveform with the form of that segment repeated every T_{segm}. This will be an accurate representation of the periodic waveform if $T_{\text{segm}} = T_{\text{sequ}}$ (Fig. S.1) or if $T_{\text{segm}} = nT_{\text{sequ}}$ (Fig. S.2), where n is an integer. A Fourier transform of a waveform segment of duration T_{segm} will have frequencies spaced by cycle/T_{segm}, but if $n > 1$, it follows that only one of each n of those will be nonzero, since a waveform that repeats at intervals of T_{sequ} contains only harmonics of cycle/T_{sequ} (Fig. S.2, right side).

Advanced Frequency Synthesis by Phase Lock, First Edition. William F. Egan.
© 2011 John Wiley & Sons, Inc. Published 2011 by John Wiley & Sons, Inc.

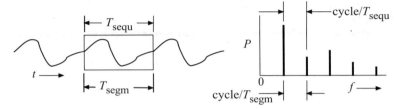

FIGURE S.1 An accurate power spectrum is obtained when the analyzed segment T_{segm} equals the sequence duration T_{sequ}.

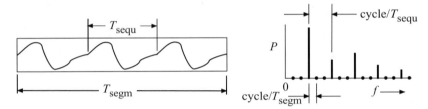

FIGURE S.2 When $T_{\text{segm}} = nT_{\text{sequ}}$, where n is an integer, power appears only at every nth DFT frequency.

S.1.2 Accurate Representation

A DFT of a waveform segment is the Fourier series of samples of the waveform at the DFT sampling rate, f_s. It produces sinusoids at frequencies that are harmonics of

$$\text{cycle}/T_{\text{segm}} = f_s/L_{\text{segm}}$$

replicated at frequency separations of f_s (i.e., at frequency offsets that are multiples of f_s), where L_{segm} is the segment length (the buffer size, the number of stored values). For example, if we sample a waveform at $f_s = 1\,\text{MHz}$ and put the data into a buffer of size $L_{\text{segm}} = 1024$, the duration of the sample will be

$$T_{\text{segm}} = 1024\,\text{cycle}/1\,\text{MHz} = 1024\,\mu s$$

and the DFT will contain harmonics of

$$\text{cycle}/1024\,\mu s = 1\,\text{MHz}/1024 \approx 977\,\text{Hz}.$$

To be an accurate representation of the repetitive waveform, again $T_{\text{segm}} = nT_{\text{sequ}}$, so

$$L_{\text{segm}} = f_s T_{\text{segm}}/\text{cycle} = nT_{\text{sequ}}f_s/\text{cycle} = nL_{\text{sequ}}f_s/f_{\text{ref}},$$

where f_{ref} is the frequency at which the sequence occurs. For example, if the sequence length is $L_{\text{sequ}} = 1024$ and $f_{\text{ref}} = 10\,\text{MHz}$, the sequence will last for

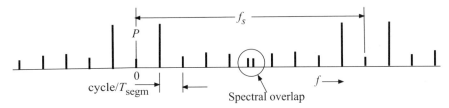

FIGURE S.3 Spectrums overlap if the sampling frequency f_s is not high enough for the spectral bandwidth.

$$T_{sequ} = 1024 \, \text{cycle}/10 \, \text{MHz} = 102.4 \, \mu\text{s}.$$

If the waveform is sampled at 2 MHz, the buffer for collected samples should then have the length

$$L_{segm} = n(1024)2 \, \text{MHz}/10 \, \text{MHz} = n \, 1024/5.$$

A choice of $n = 5$ would give a binary number for the buffer length. Then the segment would contain five repetition cycles of the sequence.

The sampling frequency f_s must also be high enough to prevent the spectrums from overlapping significantly; that is, it must be high compared to the frequencies that are in the unsampled sequence, the sequence prior to sampling (Fig. S.3). The unsampled sequence is generally low-pass filtered before sampling to ensure that this occurs, but the filtering distorts the original sequence to the degree that the sequence has energy at frequencies not passed by the filter.

S.1.3 Approximate Representation

If $T_{segm} \neq nT_{sequ}$, the DFT will represent the waveform with a different periodicity than the actual waveform (see Fig. S.4; leakage will be discussed in Section S.3). If $T_{segm} = kT_{sequ}$, $k < 1$, the DFT represents a portion, of the waveform, that is kT_{sequ} long and has a periodicity of kT_{sequ}.

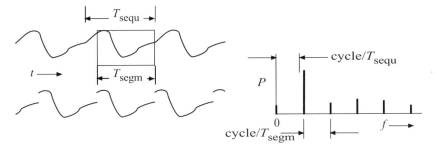

FIGURE S.4 When T_{segm} is not a multiple of T_{sequ}, the frequencies of spectral lines in the DFT are offset from the frequencies in the waveform.

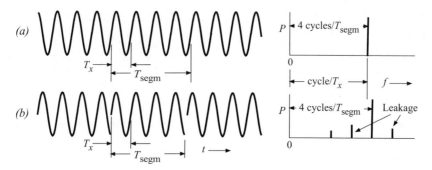

FIGURE S.5   When there are many periods of a sinusoid in T_{segm}, and $T_{segm} = nT_x$, n being an integer, the frequency of the line representing the sinusoid will be accurate (a). Otherwise, it will be offset somewhat in frequency (b) and other lines will be produced due to the finite responses in multiple bins.

If the waveform contains a component with periodicity $T_x \ll T_{segm}$, the DFT will have a corresponding value at a *multiple* of $f_1 = cycle/T_{segm}$ near $f_x = cycle/T_x$ (Fig. S.5). The DFT will be maximum at the frequency nf_1 that is closest to f_x. The maximum frequency error is $cycle/(2T_{segm})$, so the more cycles of the waveform that are contained in T_{segm}, the smaller will be the relative frequency error,[S1] which is $1/(2n)$. Thus, the DFT can give a good representation of discrete spurs in the waveform, limited by the restriction that it can only show responses at harmonics of $cycle/T_{segm}$.

However, variations in the waveform with longer periods produce different results in subsequent DFTs because the waveforms existing during each T_{segm} are different. This can result in a variation of the low-level components, which creates an apparent variation in the PSD, from DFT to DFT, in a manner similar to what would be produced by noise (as opposed to the deterministic sequence).

S.1.4 Representation of a Sequence

Here, we refer to the numerical sequence generated by the $\Sigma\Delta$ modulator rather than to a voltage or a phase produced as a result of that sequence. The transfer function of the $\Sigma\Delta$ modulator is initially given in terms of z-transforms and represents the sampled response to a sampled input; that is, the output [Eq. (M.13)] exists only at sample instants. The spectrum is replicated at the sample frequency f_{ref}. From these samples, the resulting phase is computed at sample instants [Eq. (Q.6)]. The output of a PFD is approximated as narrow pulses with areas proportional to these sampled values, while the output of a S&H PD is proportional to these values held in a zero-order hold.

A DFT that uses each number in the sequence from a $\Sigma\Delta$ modulator is the DFT of the sampled output, the values of which exist only at sample instants. Here, $f_s = f_{ref}$ and there is no filtering. This is the process that is used in the "SD sequence" analyzer and the "PPSD from sequence" analyzer, which produced Fig. 5.14, and the process that produced Fig. 6.13. The digital waveforms in the $\Sigma\Delta$ modulator are constant between clocks. If they were analyzed as in analog signals, their spectrums would be affected by the zero-order hold transfer function, $sinc(f/f_{ref})$, and by the filtering in the analyzer.

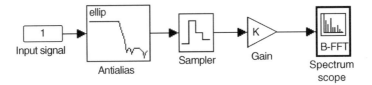

FIGURE S.6 Block diagram of a spectrum analyzer.

S.2 THE SPECTRUM ANALYZER

For each spectrum analyzer (Fig. S.6), we specify the bandwidth B of a low-pass input filter. The output of the filter is sampled at a rate

$$f_s = 2.56B, \tag{S.1}$$

and the samples are then amplified and stored in a buffer of size $L_{\text{buf}} = 2^{n_{\text{buf}}}$ within the Spectrum Scope. The DFT is developed from the data in the buffer. The number of frequency bins across f_s is L_{buf}, so each bin has width

$$\Delta f_c = \frac{f_s}{L_{\text{buf}}} = \frac{2.56B}{L_{\text{buf}}}. \tag{S.2}$$

S.3 THE WINDOW FUNCTION

The data are multiplied by a Hann window function (Fig. S.7), giving each bin a noise bandwidth[S2]

$$B_n = 1.5\Delta f_c = \frac{3.84B}{L_{\text{buf}}}. \tag{S.3}$$

Any spectral line of power P_c that falls at the center of a bin will produce a maximum response. It will also produce a 6 dB weaker response in the two adjacent bins but no response in other bins. In other words, the adjacent bins will have the same window pattern but will be shifted by one bin width (0.01 in Fig. S.6), so they will have 6 dB weaker responses. (Do not be concerned about the absolute value of the response; it will be adjusted for our purposes.) Other bins are shifted by multiple bin widths, so their response will be null.

However, if the spectral line should be offset by half a bin width, the response would be reduced by 1.4 dB (scalloping loss) and would appear at that amplitude in two adjacent bins and at reduced amplitudes in many other bins further from the center, making the line appear broader. This is called *leakage*. For lesser offsets, the closer the spectral line is to the center of a bin, the narrower the spectrum appears.

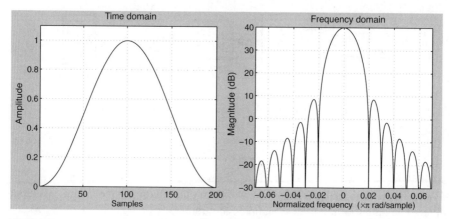

FIGURE S.7   200-bin Hann window, from MATLAB tool `wvtool(hann(200))`, buffer weighting on the left and part of frequency response on the right.

Leakage can be a problem, especially if we are looking at low levels of PSD close to the main spectral line, because leakage from that powerful line can block the PSD we are trying to observe. The solution is to ensure that the main spectral line is centered in a bin.

If no window were purposely employed, a rectangular window would nevertheless occur (all buffer elements equally weighted) and the responses farther from the spectral line would be greater; that is why the Hann window is employed.

S.4   DENSITY AND DISCRETE SPURS

The same response that is produced by a spectral line of power P_c at the center of a bin could also be produced by a PSD of $S = P_c/B_n$ extending over all the bins. We could calibrate for discrete power, but we will calibrate for density. If we know that a response is produced by a discrete signal (we can usually tell from the response), we must multiply the indicated PSD by B_n to find the power of that signal.

The spectrum of the phase in the $\Sigma\Delta$ synthesizer, after it has settled, is periodic, with period T_{sequ}, so all spectral lines are discrete. (This assumes no actual noise, which we are not simulating.) However, when $B_n \gg 1/T_{\text{sequ}}$, there are many lines in each bin, and we treat their power as a power spectral density S. If, however, the contribution of a single line is large enough, it will noticeably raise the PSD at the bin in which it exists and we will recognize its presence and call it a discrete spur. For the spur to be seen, its power P_{spur} must make a significant difference in the observed spectrum (Fig. S.8), requiring

$$P_{\text{spur}} > SB_n. \tag{S.4}$$

Therefore, what we recognize as a spur depends on B_n. Even if we should measure each of the spectral lines, rather than treating them as a density, we would have to

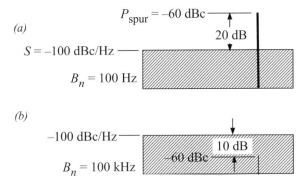

FIGURE S.8   Spur can be seen above noise in (*a*), but not in (*b*), where B_n is wider.

define what power level we consider to be a discrete spur. A potential problem arises because we are, perhaps without realizing it, making that definition when we choose a noise bandwidth. In practice, the definition must be established based on system requirements. That is, it is not adequate to specify a density without defining the noise bandwidth in which it is measured.

S.5 CONTROL PARAMETERS

There is a "Buffer overlap" parameter that causes the transform to be formed before the entire buffer is filled with new data. We generally multiply L_{buf} by an `overlap` of 0.75, which causes a new transform to be formed (a frame to occur) whenever 25% of L_{buf} is new data (75% of the data are retained from the previous frame). Thus, the data are completely changed every four frames and we see updates four times faster than we would if we waited for the whole buffer to fill. We also generally show `averages` of the last four plots. In addition, the buffer size is specified as

$$L_{buf} = L_{buf0}/\texttt{crs}, \qquad\qquad (S.5)$$

where L_{buf0} is a constant and `crs` is a model-wide parameter for adjusting the coarseness of the transforms. If `crs` is set to 4, the transforms will be computed four times faster, but B_n will be four times wider, and conversely. The parameters `overlap` and `averages` are also model wide, so the response of all of the spectrum analyzers can be adjusted at once.

S.6 FREQUENCY CONVERSION IN AN ANALYZER

The spectrum analyzers generally display baseband, starting at zero frequency, but one spectrum analyzer mixes the synthesizer output signal with a fixed oscillator to center the resulting signal at half of the specified bandwidth. Reasons for using frequency conversion are discussed in Section 6.1.8.

S.7 DISPLAYING L, FPSD, AND PPSD

As a starting point, we have set the spectrum analyzers to display one-sided PSD in dB relative to 1 W/Hz. If we wish to display L (top two SAs), the signal must be amplified to unity power, so the PSD will also equal the relative (to 1) PSD. Then the PSD of the sidebands will be relative to the power of the signal. The VCO in our models produces a sinusoid with unity amplitude (imagine 1 V if that is helpful), corresponding to a power of 0.5 (0.5 W in a 1 Ω load). Therefore, the total voltage gain preceding the Spectrum Scope in order for it to display L must be[S3]

$$K = \sqrt{2}. \tag{S.6}$$

We have chosen to show two-sided FSPD S_{2f} (two middle SAs), so we can divide it by f_m^2 to produce $\mathcal{L}$. In order to show S_{2f}, the tuning voltage is multiplied by the tuning sensitivity (converting voltage to frequency deviation) and divided by $\sqrt{2}$ (halving the PSD), requiring a gain before the Spectrum Scope of

$$K = \mathrm{sen}/\sqrt{2}. \tag{S.7}$$

Here, sen is a model-wide parameter equal to K_v.

We wish to show the theoretical $S_{2\varphi} = \mathcal{L}$ due to quantization noise (referenced to the synthesizer output) in units of dBr/Hz (lowest SA), requiring a gain preceding the Spectrum Scope of

$$K = 2\pi/\sqrt{2}. \tag{S.8}$$

This converts the numerical sequence n_d, which represents time shift in units of T_{out}, to radians of phase at F_{out} and divides the one-sided PSD by 2. We can show the two-sided PSD of the numerical sequence itself (second lowest SA) by just letting K equal $1/\sqrt{2}$.

S.8 SPECTRAL OVERLAPS

S.8.1 Aliasing

We can set the bandwidth B of an SA by double-clicking on its icon. The bandwidth of the antialias filter (Fig. S.6) will be set to B and the sample frequency f_s will be set to $2.56B$, producing the condition illustrated in Fig. S.9. The illustration of the passband at a_0 represents the response to flat noise or to a constant signal amplitude versus frequency. The responses at $a_{\pm 1}$ represent the closest aliased responses. The guard band B_g has width $0.56B$. The minimum attenuation A of aliased signals inside the main response band is the attenuation of the filter at a shape factor of 1.56. For most of the SAs used in our simulations, the attenuation there is 70 dB. In many cases, the

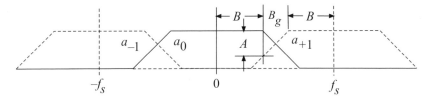

FIGURE S.9   SA filter responses to a flat spectrum. Main response is a_0; $a_{\pm 1}$ are aliased responses.

signal being observed will also be attenuated significantly at 1.56B, further reducing interference.

S.8.2 Spectral Folding

The synthesizer's output spectrum is often observed after frequency conversion, as discussed in Section S.6. The output is frequency converted to $B/2$ in the SA. The resulting spectrum is illustrated in Fig. S.10. Above $B/2$ from spectral center, the output will be attenuated by the low-pass filter in the SA. Equidistant on the low side, the output will receive interference from folding of the spectrum.

The signal centered at F_{out} has been converted to F_{SA} by mixing with

$$F_{LO} = F_{out} - B/2. \tag{S.9}$$

Output frequencies at

$$f = F_{out} - B/2 - \Delta f, \quad \Delta f > 0, \tag{S.10}$$

will be negative after conversion and will be seen at the spectrum analyzer at a frequency

$$f_{SA} = |f - F_{LO}| = \Delta f \tag{S.11}$$

on top of those originally at

$$f = F_{out} - B/2 + \Delta f, \tag{S.12}$$

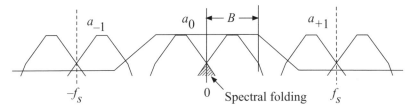

FIGURE S.10   Converted spectrums showing interference from image and attenuation due to filter.

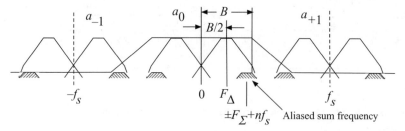

FIGURE S.11   Converted spectrums showing interference from aliased spectrums.

thus distorting the observed spectrum. Therefore, it is important that the power in the observed spectrum be small by $B/2$ from spectral center. This limits how small we can choose B and shows why the elliptic filter in the loop filter is important; it permits us to better observe the synthesizer output.

S.8.3 Image

In addition to the desired difference frequency f_Δ, the mixing process produces an image frequency at

$$f_\Sigma = f_{\text{out}} + F_{\text{LO}}. \tag{S.13}$$

After reduction by the ultimate attenuation of the SA low-pass filter, sampling will cause signals at these frequencies to be aliased [Lyons, 2004, Chapter 2] to

$$f_{\text{SA}} = f_\Sigma + nf_s, \tag{S.14}$$

where n is any positive or negative integer, and appear at $|f_\Sigma + nf_s|$ on top of the true spectral picture in the region (Fig. S.11). The filter used in the Converter & SA has an ultimate attenuation of $120\,\text{dB}$.

S.9 ANOMALOUS SPURS

If we remove all modulation from the Simulink model, we still see spurs and broadband noise on the SA displays at some level, depending on the parameters used. These anomalous spurs have been observed to disappear when a fixed-step solver is used. However, a variable-step solver is appropriate for the overall simulation. If there is a question about the source of certain observed spurs, it may be desirable to see whether they occur in steady state and in the absence of $\Sigma\Delta$ modulation. However, they do not appear to be a practical problem in the models we discuss. The solver and related parameters are specified in the Configuration Window (`Simulation/Configuration Parameters...` or `Control-E`). See the Simulink help documents regarding the solvers.

APPENDIX T

TOOLBOXES

In this appendix, we list the toolboxes required for MATLAB scripts and the block sets required for Simulink models.

Thirteen scripts require the Signal Processing Toolbox. They are `ct_.m`, `sin_.m`, `closed_.m`, `error_.m`, and `open_.m`, where _ represents any group of symbols. Only `SynCP1012`, which is from *FS2*, requires the Control System Toolbox. The other nine require only the MATLAB program.

Table T.1 shows the block sets required to run the various models. Models that contain circuit components require SimPowerSystems.

The MATLAB scripts and Simulink models will run on the Student Version, but the accelerator is not available in that version, so simulations will be slower.

Advanced Frequency Synthesis by Phase Lock, First Edition. William F. Egan.
© 2011 John Wiley & Sons, Inc. Published 2011 by John Wiley & Sons, Inc.

TABLE T.1 Mathworks Products Required (+) for the Models

PRODUCTS	SimplePD	CPandI CPandIfloor CPandIplus Dither EFeedback FeedForward HandK HandKsimple SandH SandHreverse	SynStepSH2 SynStepSH3 SynStepSH4	SynStepSH* FFmodTest* FBmodTest
MATLAB©	+	+	+	+
Simulink©	+	+	+	+
Communications Blockset	+	+	+	SynStepSH +
Signal Processing Blockset	+	+		FFmodTest +
SimPowerSystems	+		+	

*See one additional requirement in column below for SynStepSH and FFmodTest.

APPENDIX U

NOISE PRODUCED BY CHARGE PUMP CURRENT UNBALANCE (MISMATCH)

Here, we compute the noise floor produced as a result of mismatch between the two charge pump currents in the presence of $\Sigma\Delta$ modulation (the development is similar to that in Arora et al. [2005]).

If the currents in a charge pump are mismatched, such that one exceeds the other by a small factor k_u, we can look on the process as one where a sequence of positive and negative pulses that causes the ideal quantization noise spectrum is now accompanied by a second mismatch error (*u*-error) sequence consisting of $k_u/2$ times the (say) negative pulses in that sequence and $-k_u/2$ times the positive pulses, as shown in Fig. U.1. Here, the "detected" waveform represents the sample-and-hold value of the phase measured by the PD.[U1] [In a type-2 loop, the whole detected waveform might become slightly higher in order to produce zero average PD output, but we are now not interested in the resulting DC term (assuming it is small enough[U2]).] We expect the narrow charge pump pulses to have the same low-frequency content as the sample-and-hold waveform so that the low-frequency noise produced by the mismatch will be the same. This is the same assumption that is made when we represent phase detection as a linear continuous process, as in Fig. F.1.19*b*.

The *u*-error sequence would have an amplitude equal to $k_u/2$ times the amplitude of the rectified ideal sequence. Its variance σ_u^2 would therefore be $k_u^2/4$ times the variance of the latter,

$$\sigma_u^2 = \frac{k_u^2}{4} r_q \sigma_q^2,$$ (U.1)

Advanced Frequency Synthesis by Phase Lock, First Edition. William F. Egan.
© 2011 John Wiley & Sons, Inc. Published 2011 by John Wiley & Sons, Inc.

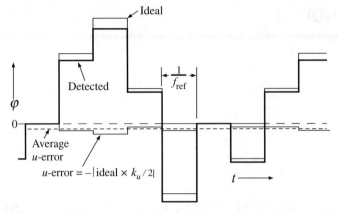

FIGURE U.1 Detected and ideal phase waveforms. Mismatched charge pump creates waveform that can be considered the sum of an ideal waveform containing quantization noise and an error (u-error) waveform.

where σ_q^2 is the variance of the ideal (theoretical) quantization noise and r_q is the ratio of the variance of the rectified sequence to that of the original sequence.

On the basis of Eq. (2.3), the integrated PPSD for the quantization noise for pth-order MASH–11... is

$$\sigma_{qp}^2 = \int_0^{f_{ref}/2} S_\varphi(f_m) df_m = \frac{(2\pi \text{ rad})^2}{6 f_{ref}} 2^{2(p-1)} \int_0^{f_{ref}/2} \sin^{2(p-1)}\left(\pi \frac{f_m}{f_{ref}}\right) df_m \qquad (\text{U.2})$$

$$= \frac{(2\pi \text{ rad})^2}{6 f_{ref}} 2^{2(p-1)} \frac{f_{ref}}{\pi} \int_0^{\pi/2} \sin^{2(p-1)} x \, dx \qquad (\text{U.3})$$

$$= \frac{(2\pi \text{ rad})^2}{6\pi} 2^{2(p-1)} \frac{1 \cdot 3 \cdot 5 \cdots (2p-3)}{2 \cdot 4 \cdot 6 \cdots (2p-2)} \frac{\pi}{2} \qquad (\text{U.4})$$

$$= \frac{(2\pi \text{ rad})^2}{3} 2^{(2p-4)} \frac{1 \cdot 3 \cdot 5 \cdots (2p-3)}{2 \cdot 4 \cdot 6 \cdots (2p-2)}. \qquad (\text{U.5})$$

Combining Eqs. (U.1) and (U.5), we obtain

$$\sigma_{up}^2 = k_u^2 r_q \frac{(2\pi \text{ rad})^2}{3} 2^{(2p-6)} \frac{1 \cdot 3 \cdot 5 \cdots (2p-3)}{2 \cdot 4 \cdot 6 \cdots (2p-2)}. \qquad (\text{U.6})$$

If the noise power due to the mismatch is spread uniformly over frequency, this leads (under the assumption of small modulation index) to a single-sideband relative noise power density, owing to mismatch, of

$$\mathcal{L}_{up}(k_u) = \frac{\sigma_{up}^2(k_u)}{f_{ref}} \, \text{rad}^2, \tag{U.7}$$

where "(k_u)" indicates a functional dependence. Here, we have divided the phase power σ_{up}^2 by $f_{ref}/2$, to get a phase density (see Section N.4), and then by 2 to get $\mathcal{L}_{up}$. Expanding, we obtain

$$\mathcal{L}_{up}(k_u) = \frac{k_u^2}{f_{ref}} r_{qp} \frac{(2\pi)^2}{3} 2^{(2p-6)} \frac{1 \cdot 3 \cdot 5 \cdots (2p-3)}{2 \cdot 4 \cdot 6 \cdots (2p-2)}. \tag{U.8}$$

The value of r_{qp} was determined by analysis of sequences from a $\Sigma\Delta$ modulator of order $(p-1)^{U3}$ (see Section 6.7.3). Results are shown in Fig. U.2. From these, we set the following values for r_{qp}:

$$p = 3, \quad r_{q3} = 0.5 + 0.5 n_{fract},$$
$$p = 4, \quad r_{q4} = 0.38 + 0.19 n_{fract},$$
$$p = 5, \quad r_{q5} = 0.34 + 0.07 n_{fract},$$
$$p > 5, \quad r_{qp>5} = 0.34.$$

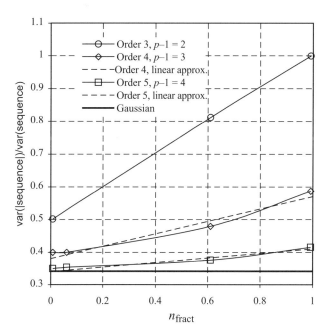

FIGURE U.2 Ratio of variance of absolute value to variance of sequence as a function of the divider fraction. Each data point represents the average of the statistics from pairs of 10,000-point samples.

The last line uses r_q for a Gaussian [Davenport and Root, 1958, Eq. (12.92)], which the curves appear to be approaching.

For the various values of p, Eq. (U.8) becomes

$$\mathcal{L}_{u3}(k_u) = \frac{k_u^2}{f_{\text{ref}}} r_{q3} \frac{(2\pi)^2}{8} \approx \frac{k_u^2}{f_{\text{ref}}} 0.5(1 + n_{\text{fract}}) \frac{(2\pi)^2}{8} = \frac{k_u^2}{f_{\text{ref}}}(1 + n_{\text{fract}})2.5, \quad (\text{U.9})$$

$$\mathcal{L}_{u4}(k_u) = \frac{k_u^2}{f_{\text{ref}}} r_{q4} \frac{(2\pi)^2}{8} \frac{10}{3} \approx \frac{k_u^2}{f_{\text{ref}}}(1 + 0.5n_{\text{fract}})6.25, \quad (\text{U.10})$$

and

$$\mathcal{L}_{u5}(k_u) = \frac{k_u^2}{f_{\text{ref}}} r_{q5} \frac{(2\pi)^2}{8} \frac{35}{3} \approx \frac{k_u^2}{f_{\text{ref}}}(1 + 0.2n_{\text{fract}})19.6. \quad (\text{U.11})$$

For $f_{\text{ref}} = 10\,\text{MHz}$ and $k_u = 0.01$, these are

$$-106\,\text{dBc/Hz} \leq \mathcal{L}_{u3} \leq -103\,\text{dBc/Hz}, \quad (\text{U.12})$$

$$-102\,\text{dBc/Hz} \leq \mathcal{L}_{u4} \leq -100.3\,\text{dBc/Hz}, \quad (\text{U.13})$$

$$-97.1 \leq \mathcal{L}_{u5} \leq -96.3\,\text{dBc/Hz}. \quad (\text{U.14})$$

Arora et al. [2005] obtain values close to these for MASH-111 and MASH-1111, and also obtain values for many more configurations.[U4]

Of course, when $n_{\text{fract}} = 0$, the noise will not be present if the $\Sigma\Delta$ modulation is turned off or if the accumulators do not have an initial value.

APPENDIX W

GETTING FILES FROM THE WILEY INTERNET SITE

To view the files listed in this book and other materials associated with it, please go to www.wiley.com/go/EganAdvanced.

Advanced Frequency Synthesis by Phase Lock, First Edition. William F. Egan.
© 2011 John Wiley & Sons, Inc. Published 2011 by John Wiley & Sons, Inc.

APPENDIX X

SOME TABLES

X.1 ACCUMULATOR SHORTENING

The number of bits in each accumulator is in the first column of Table X.1 and the required value of a_{HK} (reduction in modulus) is in the second column [Hosseini and Kennedy, 2007, Table II]. The resulting accumulator lengths are shown in Table X.2.

TABLE X.1 Value of a_{HK}

n_0	a
5, 7, 13, 17, 19	1
6, 9, 10, 12, 14, 20, 22, 24	3
8, 18, 25	5
11, 21	9
16, 23	15
15	19

Advanced Frequency Synthesis by Phase Lock, First Edition. William F. Egan.
© 2011 John Wiley & Sons, Inc. Published 2011 by John Wiley & Sons, Inc.

TABLE X.2 Resulting Accumulator Lengths

Accumulator Length	n_0	a
31	5	1
61	6	3
127	7	1
251	8	5
509	9	3
1021	10	3
2039	11	9
4093	12	3
8191	13	1
16,381	14	3
32,749	15	19
65,521	16	15
131,071	17	1
262,139	18	5
524,287	19	1
1,048,573	20	3
2,097,143	21	9
4,194,301	22	3
8,388,593	23	15
16,777,213	24	3
33,554,427	25	5

X.2 SEQUENCE LENGTHS

The sequence length for n-bit accumulators in a p-section MASH-111... configuration is 2^m, where m is given in Table X.3 [Borkowski et al., 2005, Table II], if the initial condition is odd (i.e., LSB is set in the first accumulator).

This also applies to fractions containing a small LSB, 2^{-n}, with no initial setting, since the register (accumulator) is then equivalent to an n-bit register with the LSB set after the first clock. In this case, $m \geq n$, since $m = n$ for the first register. None of the bits that are less significant than the nth LSB participate in the operation of the registers under these conditions.

TABLE X.3 Minimum and Maximum Values of m

p	Minimum m	Maximum m
2	$n - 1$	$n + 1$
3	$n + 1$	$n + 1$
4	$n + 1$	$n + 2$
5	$n + 2$	$n + 2$

END NOTES

CHAPTER 2

1. Meninger and Perrott, 2003.
2. Pamarti et al., 2004.
3. An extensive discussion of the amplitude of fractional spurs is contained in Banerjee [2006, Chapter 12].
4. This is digital differentiation. King [1980] uses a small capacitor to differentiate the value in the second accumulator after DA conversion.
5. $8192/20.48 \times 10^6 = 4 \times 10^{-4}$.
6. The sample rate was changed by changing the specified SA bandwidth. Since the spectrum is converted to half of the SA bandwidth, the spectral center has shifted in Fig. 2.15.
7. These analyzers are set to make a new frame when one-fourth of the buffer has been filled with new data, so T_{segm} is four times longer than the time to create a new frame.
8. $B_n =$ (window factor)(sample frequency)/(buffer size), (window factor) $= B_n$/(cell width) $= 1.5$ for Hann window, (sample frequency) $= 2.56$(spectrum analyzer bandwidth) $= 2.56(8\,\text{MHz})$ for Fig. 2.14. $B_n = 1.5(2.56)(8\,\text{MHz})/8192 = 3750\,\text{Hz}$ for Fig. 2.14.
9. MASH stands for multistage noise shaping and MASH-*abc* indicates that the stages have orders *a*, *b*, and *c*. That is, the first stage has *a* poles in its transfer function $F(z)$, and so on.
10. Third-order MASH: Crawford, 2008, pp. 355–358; De Muer and Steyaert, 2003, pp. 174–175; Kozak and Kale, 2004, pp. 1148–1162; Kroupa, 2003, pp. 290–301; Lacaita et al., 2007, pp. 62–63; Pamarti and Galton, 2007; Rogers et al., 2006, pp. 319–325; Shu et al., 2001; Shu and Sánchez-Sinencio, 2005, pp. 74–76, 87–88.

Advanced Frequency Synthesis by Phase Lock, First Edition. William F. Egan.
© 2011 John Wiley & Sons, Inc. Published 2011 by John Wiley & Sons, Inc.

11. US Patent 7,315,601.
12. We note that the center of the pulses from the PFD varies from the reference instant by half of the (time equivalent to the) phase change. With a S&H PD, the center of the level change varies as the average of two adjacent phase changes, the one that initiates the level change and the one that terminates it. Moreover, the duration of that level change varies. Thus, the varying sample rate not only produces a time modulation of the S&H PD output but also affects the average value of each level change. With the PFD, in contrast, the average value of the pulse is not affected by the varying sample rate.
13. (Sample frequency)/(buffer size) $= (2.56 \times 8\,\text{MHz})/8192 = 2500\,\text{Hz}$.
14. Filiol et al. [1998] were using MASH-1111. Kozak and Kale [2004] found that a MASH-111 modulator could produce shaped white noise, whereas there was some autocorrelation for MASH-11. Although the phase noise in the loop is accumulated, producing a reduction of one order in the spectral shape relative to the output from the modulator, it seems unlikely that the accumulation would affect the existence, or lack of, discrete spurs.

CHAPTER 3

15. Filiol et al. [1998] and Kozak and Kale [2004], citing them, set the LSB each time n_{fract} is changed. A value of n_{fract} that contains an LSB can reset the LSB in the modulator (make it equal to zero) and could leave the final value with other small digits reset also, thus not providing a good initial value for the next input.
16. We use $\mathcal{L}_\varphi$ rather than L_φ because the variance of the integrated phase noise is not small compared to a radian (Section 1.5). However, $\mathcal{L}_\varphi$ will become equal to L_φ once the loop response is accounted for.
17. The spectral pictures vary some with each frame (Section S.1.3).
18. $2^{n-1} + 2^{n-1} + 1 = 2^n + 1 < 2^{n+1}$.

CHAPTER 4

19. $(10\,\text{dBc/Hz}) \log_{10}(k_1\text{Hz})$ is the figure of merit used by Banerjee and others for white noise. Banerjee calls it *PN1Hz*. *PN10kHz* $= (10\,\text{dBc/Hz}) \log_{10}(k_{11}\text{Hz}) + 140\,\text{dB}$ is the constant used by Banerjee for flicker noise; it equals the flicker noise at $f_m = 10\,\text{kHz}$ and $f_{\text{out}} = 1\,\text{GHz}$. We have also substituted $I_{\text{knee}}/I_{\text{cp}}$ for his $K_\phi Knee/K_\phi$, since $K_\varphi \sim I_{\text{cp}}$. Banerjee provides tables giving the important parameters for many ICs. He also notes that the flicker noise sometimes discontinues its climb as the frequency is reduced and flattens out at low frequencies. Without knowing circuit details, the reason for this is difficult to determine.
20. Since we will plot noise above 1 kHz f_m, incorporating a break just above that frequency would add complexity without any appreciable benefit.
21. These were obtained using the spreadsheet IntPhNs.xls.
22. So the resulting noise spectrum will be wide.
23. The approximation is best for $w_{\text{nb}} \gg f_{\text{ref}}$.
24. Arora et al. [2005] indicate this explicitly for the second and third effects. For the first, they give that the noise current is proportional to $1/T_{\text{ref}}$. The noise power, the density of which is

observed to be white, would be spread over $f_{ref}/2$ (Section N.4), making the power density proportional to $1/(T_{ref}^2 f_{ref}) = f_{ref}$.

25. This assumes small modulation index, which would be usual.

26. Based on its frequency dependency, this may be what Banerjee [2006] refers to as the "pulse spur."

27. From Eq. (M.32) and Section M.5.

28. From Eq. (M.30) and Section M.5.

29. De Muer and Steyaert [2003], Fig. 6.21, at 600 kHz offset.

CHAPTER 5

30. Crawford, 2008, pp. 351–355; Lacaita et al., 2007, pp. 62–63; Lee et al., 2001b; Rogers et al., 2006, pp. 326–330; Shu et al., 2001; Shu and Sánchez-Sinencio, 2005, pp. 84–88.

31. 14 bit registers, initially 0 except LSB set in the first register. 2^{15} clock cycles for each of 400 n_{fract} values spread evenly from 0 to 1.

32. Crawford, 2008, pp. 348–351; De Muer and Steyaert, 2003, pp. 175–178; Rhee et al., 2000; Riley et al., 1993; Rogers et al., 2006, pp. 338–342; Shu et al., 2001; Shu and Sánchez-Sinencio, 2005, pp. 81–84, 87–88.

33. The step response is

$$\frac{1}{1-z^{-1}} \left[\frac{2z^{-1}-2.5z^{-2}+z^{-3}}{1-z^{-1}+0.5z^{-2}} \right] = \frac{2z^{-1}-2.5z^{-2}+z^{-3}}{1-2z^{-1}+1.5z^{-2}-0.5z^{-3}}.$$

We can multiply the two denominators in the left expression using the MATLAB conv function if we wish. The denominators are represented as arrays of coefficients of z^{-n}, starting with the highest n:

$$d_1 = [-1 \quad 1]; \quad d_2 = [0.5 \quad -1 \quad 1]; \quad d = \text{conv}(d_1, d_2) = [-0.5 \quad 1.5 \quad -2 \quad 1].$$

Then, we can pad the numerator with zeros on the right and use the MATLAB deconv function to obtain the response by dividing the denominator into the padded numerator:

$$n = [1 \quad -2.5 \quad 2 \quad 0 \quad 0 \quad \cdots \quad 0]; \quad o = \text{deconv}[n,d] = [2 \quad 1.5 \quad 1 \quad 0.75 \quad 0.75 \quad 0.875 \quad 1 \quad 1.0625\cdots];$$
$$\text{error} = o - [1 \quad 1 \quad 1 \quad \cdots \quad 1] = [1 \quad 0.5 \quad 0 \quad -0.25 \quad -0.25 \quad -0.125 \quad 0 \quad \cdots].$$

34. Data were gathered from simulations at $n_{fract} = 2^{-15}$, 0.4, 0.9, and $(1-2^{-15})$. Rhee et. al. show their results at $n_{fract} = 0.25 + 2^{-7}$.

35. Rhee et. al. [2000] appear to use pseudorandom LSB dither in their IC while, instead, setting the 2^{-16} LSB in simulations that produce their figures 7 and 8, referring to both as dither.

36. Crawford, 2008, pp. 357–358; Pamarti and Galton, 2007; Rogers et al., 2006, pp. 325–326; Shu et al., 2001; Shu and Sánchez-Sinencio, 2005, pp. 76–81, 87–88.

37. Example: If $n(1) = 16$ and $n(3) = 9$, the bits dropped in third accumulator are Eq. (5.9) $\Rightarrow [0, 1, \ldots, 127]2^{-16}$.

38. Continuing example above,

$$\text{Eq. (5.10)} \Rightarrow \frac{2^{-18}-2^{-32}}{12} \doteq \frac{2^{-32}(4095)}{12} \approx \frac{2^{-32}(4096)}{12} = \frac{2^{-18}}{12} \Rightarrow \text{Eq. (5.11)}.$$

CHAPTER 6

39. These models were developed from the "fractional_6" model in the set called "ams" on MATLAB Central (http://www.mathworks.com/matlabcentral/fileexchange/1320-analog-mixed-signal-examples).

40. Nevertheless, the time may not be lost, since the final spectrum shape seems to occur in a frame number lower by 1, possibly at the same time as if the scope had opened at the start of the simulation.

41. The four vector elements are x and y for lower left corner and width and height of the window in pixels. If you do not see the screen, try setting the first two elements to 30 (i.e., $[30 \quad 30 \quad \cdots \quad \cdots]$).

42. There have been reported problems getting SAs to start when they are enclosed in enabled subsystems, as these are. However, the advantage of increased speed gained by turning off unneeded SAs seems to outweigh the inconvenience of the apparently rare occurrence of these problems. The enabling feature can be defeated if desired by looking under the mask of the subsystem and deleting the "Enable" icon.

43. AMD Athlon™ 3000+ XP processor.

CHAPTER 7

44. 1275 is minimum RF from Table 7.1; 1659 is the sum of that value and the maximum IF, also from the table.

45. $\dfrac{R_1-R_2}{R_1} = x.$ $\dfrac{R_1-R_2}{R_2} = 1-x$ to account for aliasing.

$$R_1 R_2 - R_2^2 = R_1 R_2 - R_1^2 + R_1 R_2. \qquad 0 = R_1^2 - R_1 R_2 - R_2^2.$$

$$R_1 = \frac{R_2 \pm \sqrt{R_2^2 + 4R_2^2}}{2} = 1.618 R_2. \qquad R_1 R_2 = 1.618 R_2^2 = 65,280.$$

$$R_2 = 200.863. \qquad\qquad R_1 = 1.618\, R_2 = 324.997.$$

46. For the previous design, $N_1 = 1250$ and $N_2 = 5250$ produce center frequencies of $\overline{F}_1 = 320,000\Delta$ and $\overline{F}_2 = 1,338,750\Delta$, corresponding to mean frequencies in Table 7.1. In the revised design, $N_1 = 1022$ and $N_2 = 6600$ produce center frequencies of $\overline{F}_1 = 332,150\Delta$ and $\overline{F}_2 = 1,326,600\Delta$, changes of $+3.8\%$ and -0.9%, respectively. This small change causes the required filter shape factor to change from 3.8 to 3.1.

CHAPTER 8

47. Although it is not usually important, a z-transform of the loop output represents a sequence of samples of that output, and it is possible that there be significant variations between samples (e.g., hidden oscillations), but Laplace transforms represent the waveform even between samples.

48. The sampling occurs at the divider transition because the information about the phase is delivered at that time. Even though the information becomes available then, the signal that indicates the phase may be offset from that instant. See Section F.7.1 for further consideration.

49. All the curves in Fig. F.7.24, except the "hold" curves, have the wrong sign on their abscissas (y-values). See the errata.

50. $-90°$ from the pole at 0 and $-45°$ from the filter at its corner frequency leave another $-45°$ before the loop becomes unstable due to $-180°$ excess phase at unity gain.

51. Resampling involves a delay, which produces additional phase shift, that must also be accounted for.

CHAPTER 9

52. From Sotiriadis [2010a]: $L = \dfrac{2^{n-m}}{\gcd(w, 2^{n-m})}$ (10). Since $\gcd(W, 2^{n-m}) = 2^{\text{LFZ}}$, $L = \dfrac{2^{n-m}}{2^{\text{LFZ}}} = 1$ implies $\text{LFZ} = n - m$, that is, all fractional bits are 0, or $k_{\text{fract}} = 0$. $T_p = \dfrac{T_{\text{out}}}{M}\dfrac{W}{\gcd(W, 2^{n-m})}$ if $2^{n-m} \le W < 2^n$ (24) $\Rightarrow$ $T_p = \dfrac{T_{\text{out}}}{M}\dfrac{W}{2^{\text{LFZ}}}$ if $1 \le k < M$. $T_v = T_p$ if $L > 1$ and $T_v = 2T_p$ if $L = 1$ (23) $\Rightarrow$ $f_v = 1/T_p$ if $k_{\text{fract}} \ne 0$ and $f_v = 1/(2T_p)$ if $k_{\text{fract}} = 0$.

53. From Eqs. (9.7) and (9.2), $\dfrac{\bar{f}_y}{f_v} = \dfrac{W2^{-\text{LFZ}}(1 + \delta_{0,k_{\text{fract}}})}{2k} = \dfrac{W2^{-\text{LFZ}}(1 + \delta_{0,k_{\text{fract}}})}{2W2^{m-w}} = 2^e \ge 1$, where $e = w - m - \text{LFZ} + \delta_{0,k_{\text{fract}}} - 1 \ge 0$. We can verify this for values of k_{fract} starting at 0. If we forgo trailing zeros in k_{fract}, we can write $e(k_{\text{fract}}) = w - m + \delta_{0,k_{\text{fract}}} - 1$. Then, $e(k_{\text{fract}}) = e(0) = m - m + 1 - 1 = 0$, $e(0.1_b) = (m + 1) - m + 0 - 1 = 0$, $e(0.01_b) = (m + 2) - m + 0 - 1 = 1$, $e(0.11_b) = (m + 2) - m + 0 - 1 = 1$, $e(0.001_b) = (m + 3) - m + 0 - 1 = 2, \ldots$.

54. $1024/200.125 \approx 5$.

55. This would be multiplied by a $\text{sinc}(\Delta f/F_{\text{ref}})$, a zero-order hold transfer function, to account for the fact that the measured phase is held throughout T_{ref} (which incorporates an approximation of constant sampling rate).

56. Hsu et al. [2008] -42 dBc at 3.67 GHz, Lee et al. [2009] -48 dBc at 1.7 GHz, Xu et al. [2009] -61 at 2.4 GHz, Temporiti et al. [2009] at 3 GHz.

APPENDIX A

A1. $S_\varphi = 2\mathcal{L}_\varphi = \mathcal{L}$, taking half of the noise to be phase noise.

A2. See Staszewski and Balsara, 2006, p. 25.

A3. $S_\varphi(f_m) = -170$ dBr/Hz $\Rightarrow \mathcal{L}_\varphi(f_m) + \mathcal{L}_\varphi(2F_{\text{ref}} - f_m) = 3$ dB -173 dBc/Hz (from spectral folding when $f_m > F_{\text{ref}}$) $\Rightarrow \mathcal{L} = -167$ dBc/Hz (assuming equal AM and FM noise).

A4. Possible experiment: configure the loop as type-2 and add $0.5/n_i$ (or $1.5/n_i$, $2.5/n_i$, etc.) to the phase error, causing the output phase to then alternate between whole multiples of $1/n_i$, acting as a bang-bang servo, when $n_{\text{fract}} = 0$. Determine how this change affects the response to the `test` signal.

APPENDIX E

E1. Cassia et al. [2003] omit a factor of 2 in what follows. The factor is included for theoretical reasons explained in Section N.6, but the data in Fig. 2.26 support its inclusion about as well as they support its omission, being midway between curves obtained both ways.

E2. $S_{2\varphi q\text{\_linear}}(f_m)$ over $-\frac{f_{\text{ref}}}{2} < f_m < \frac{f_{\text{ref}}}{2}$ is convolved with itself and then replicated to produce the repeated spectrums.

APPENDIX G

G1. It seems that $G(s)$ would stay real when s is on $jn\Omega_s$, where n is an integer, because of the symmetry of the poles and zeros in the s plane.

APPENDIX H

H1. See Section 3.3.2.12 Filter Stability in Egan [1998]. The same information is in Section 3.B.5, which is part of Appendix i.3.B in Egan [2007], but this is available only online at `ftp://ftp.wiley.com/public/sci_tech_med/phase_lock/`, where errata pertinent to Section 3.3.2.12 can also be found.

APPENDIX L

L1. `opent58`: Open-loop gain, total (including elliptic filter), 58 kHz bandwidth.

L2. `errort58`: Error, total, 58 kHz bandwidth.

L3. `errort58delay`: Error, total, with delay equal to half of a reference cycle.

L4. `ct58lin`: Closed-loop forward gain $H(f)$, linear frequency scale.

L5. `opent58over`: Open-loop gain, total, 58 kHz bandwidth, overdamped (the term strictly applies to second-order loops but is suggestive of greater phase margin).

APPENDIX M

M1. Their Eq. (14) gives the equivalent $\mathcal{L}(f_m)$.

M2. For an explanation of why the noise is so distributed, see Egan [1998], Section 18.M.1 or Egan [2007], Section i17.M.1. The latter is available at ftp://ftp.wiley.com/public/sci_tech_med/phase_lock (navigate to Phase-Lock 2nd Edition/MATLAB(_.M) Appendices. zip).

APPENDIX Q

Q1. Miller and Conley state that for a locked loop, $f_{out}(k) = N_{div}(k)f_{ref}$, which does not allow finite response times.

Q2. Z-transforms do not account for the signal dependence of the sampling time (*FS2*, p. 399). We do compute the signal that results from the variation in the sampling moment, but sampling is approximated as occurring at a fixed rate f_{ref}. Methods for reestablishing a uniform sampling rate are considered in Section 2.3.

Q3. Note that Eq. (M.1) is the input after sampling, the replicated input in the bracketed term of Eq. (F.7.28).

Q4. This is based on a forward gain whose frequency sensitive parts consist of a zero at 16.8 kHz, two poles at zero and a pole at 154 kHz, and a sixth-order elliptic filter with 2 dB passband ripple and 80 dB stopband attenuation.

APPENDIX S

S1. Relative error $= [cycle/(2T_{segm})]/[f_x = n \ cycle/T_{segm}] = 1/(2n)$.

S2. When few bins are used, we can observe that their width is not quite f_s/L_{buf}. The last bin before $f_s/2$ is about half of the normal width, widening of the other bins by a factor of about $1/L_{buf}$.

S3. In earlier versions of the Simulink program, prior to R2009b, the gains in Eqs. (S.6)–(S.8) also had to be divided by $\sqrt{1.28B}$.

APPENDIX U

U1. It might be obtained with a separate measuring circuit or by taking the difference between subsequent voltages [multiplying by $(1 - z^{-1})$] on the loop filter capacitor that is fed by the charge pump. A measuring circuit might comprise a charge pump that mimics the CP in the circuit and drives a capacitor that is discharged after its voltage is transferred to a hold circuit.

U2. Leakage current can cause a phase offset such that one current source has a fixed pulse width increase relative to the other. If leakage current becomes high enough, one of the current sources will not turn on for narrower pulses and, at high enough leakage levels, will not be used at all; the mismatch noise will disappear. However, this would generally entail other problems, such as noise from the current source and a large spectral content at the reference frequency.

U3. The phase error sequence for an order p synthesizer was obtained from the modulator that is used in an order $(p - 1)$ model, since the phase error is one order lower than the modulator output.

U4. Arora et al. [2005] use r_q for a Gaussian, that is, 0.34, for all orders.

REFERENCES

Andreani, P., and Sjöland, H. (2002). "Tail current noise suppression in RF CMOS." *IEEE Journal of Solid-State Circuits* (March). In Ravazi [2003], p. 292.

Arora, H., Klemmer, N., Morizio, J., and Wolf, P. (2005). "Enhanced phase noise modeling of fractional-N frequency synthesizers. " *IEEE Transactions on Circuits and Systems I: Regular Papers* 52(2) (February), 379–395.

Banerjee, D. (2006). *PLL Performance, Simulation, and Design*, 4th ed. Dog Ear Publishing, LLC. Available at http://www.national.com/analog/timing/pll_designbook. See also Banerjee [2008].

Banerjee, D. (2008). *"Fractional N frequency synthesis."* National Semiconductor Application Note 1879, Santa Clara, CA: National Semiconductor, December 10.

Bizjak, L., Da Dalt, N., Thurner, P., Nonis, R., Palestri, P., and Selmi L. (2008). "Comprehensive behavioral modeling of conventional and dual-tuning PLLs." *IEEE Transactions on Circuits and Systems I: Regular Papers* 55(6) (July), 1628–1638.

Boon, C. C., Do, M. A., Yeo, K. S., and Ma, J. G. (2005). "Fully integrated CMOS fractional-N frequency divider for wide-band mobile applications with spurs reduction." *IEEE Transactions on Circuits and Systems I: Regular Papers* 52(6) (June), 1042–1048.

Borkowski, M., Riley, T., Häkkinen, J., and Kostamovaara, J. (2005). "A practical Δ–Σ modulator design method based on periodical behavior analysis." *IEEE Transactions on Circuits and Systems II: Express Briefs* 52(10) (October), 626–630.

Bracewell, R. (1965). *The Fourier Transform and Its Applications.* New York: McGraw Hill.

Brennan, P., and Thompson, I. (2001). "Phase/frequency detector phase noise contribution in PLL frequency synthesizer." *Electronic Letters* 37(15) (July 19), 939–940.

Advanced Frequency Synthesis by Phase Lock, First Edition. William F. Egan.
© 2011 John Wiley & Sons, Inc. Published 2011 by John Wiley & Sons, Inc.

Cassia, M., Shah, P., and Bruun, E. (2003). "Analytical model and behavioral simulation approach for a $\Delta\Sigma$ fractional-N synthesizer employing a sample-hold element." *IEEE Transactions on Circuits and Systems II: Analog and Digital Signal Processing* 50(11) (November), 850–859.

Crawford, J. (2008). *Advanced Phase-Lock Techniques*. Boston, MA: Artech House.

Davenport, W., Jr., and Root, W. (1958). *In Introduction to the Theory of Random Signals and Noise*. New York: McGraw Hill.

De Muer, B., and Steyaert, M. (2003). "CMOS fractional-N synthesizers." *Design for High Spectral Purity and Monolithic Integration*. Boston, MA: Kluwer.

Egan, W. F. (ed.) (1972). *Miniature Low Noise Frequency Synthesizer...*, Report G1004 (June) (internal publication). Mountain View, CA: GTE Sylvania.

Egan, W. F. (1998). *Phase-Lock Basics*. New York: Wiley.

Egan, W. F. (2000). *Frequency Synthesis by Phase Lock*, 2nd ed. New York: Wiley. Abbreviated as *FS2*.

Egan, W. F. (2003). *Practical RF System Design*. Hoboken, NJ: Wiley.

Egan, W. F. (2007). *Phase-Lock Basics*, 2nd ed. Hoboken, NJ: Wiley. *Note*: Sections starting with "i." are available online at ftp://ftp.wiley.com/public/sci_tech_med/phase_lock.

Filiol, N., Riley, T., Plett, C., and Copeland, M. (1998). "An agile ISM band frequency synthesizer with built-in GMSK data modulation." *IEEE Journal of Solid-State Circuits* 33 (7) (July), 998–1008.

FS2: see Egan [2000].

Franklin, G., Powell J., and Workman M. (1990). *Digital Control of Dynamic Systems*. Reading, MA: Addison-Wesley.

Gardner, F. M. (1980). "Charge-pump phase-lock loops." *IEEE Transactions on Communications* COM-28(November), 1849–1858. Reprinted in Lindsey and Chie [1986] and Razavi [1996].

Gardner, F. M. (2005). *Phaselock Techniques*, 3rd ed. Hoboken, NJ: Wiley.

Hill, C. E. (1997). "All digital fractional-N synthesizer for high resolution phase locked loops." *Applied Microwave & Wireless* (November/December), 62–69.

Hosseini, K., and Kennedy, M. (2006). "Mathematical analysis of digital MASH delta-sigma modulators for fractional-N frequency synthesizers." *Research in Microelectronics and Electronics 2006, Ph. D.* (available in IEEE Xplore), pp. 309–312.

Hosseini, K., and Kennedy, M. (2007). "Maximum sequence length MASH digital delta-sigma modulators." *IEEE Transactions on Circuits and Systems I: Regular Papers* 54(12) (December), 2628–2638.

Hosseini, K., and Kennedy, M. (2008). "Architectures for maximum-sequence-length digital delta-sigma modulators." *IEEE Transactions on Circuits and Systems II: Express Briefs* 55 (11) (November), 1104–1108.

Hsu, C., Straayer, M., and Perrott, M. (2008). "A low-noise wide-BW 3.6-GHz digital $\Sigma\Delta$ fractional-N frequency synthesizer with a noise-shaping time-to-digital converter and quantization noise cancellation." *IEEE Journal of Solid-State Circuits* 43(12) (December), 2776–2786.

King N. (1980). "Frequency synthesizers." UK Patent Application GB 2,026,268A (January 30).

Kozak, M., and Kale, İ. (2003). *Oversampled Delta-Sigma Modulators*. Boston, MA: Kluwer.

Kozak, M., and Kale, İ. (2004). "Rigorous analysis of delta-sigma modulators for fractional-N PLL frequency synthesis." *IEEE Transactions on Circuits and Systems I: Regular Papers* 51(6) (June), 1148–1162.

Kroupa, V. (2003). *Phase Lock Loops and Frequency Synthesis.* West Sussex, England: Wiley.

Lacaita, A., Levantino, S., and Samori, C. (2007). *Integrated Frequency Synthesizers for Wireless Systems.* Cambridge: Cambridge University Press.

Lee, K., Park, J., Lee, J.-W., Lee, S.-W., Huh, H. K., Jeong, D.-K., and Kim, W. (2001a). "A single-chip 2.4-GHz direct-conversion CMOS receiver for wireless local loop using multiphase reduced frequency conversion technique." *IEEE Journal of Solid-State Circuits* 36(5) (May), 800–809.

Lee, S., Yoh, M., Lee, J., and Ryu, I. (2001b). "A 17 mW, 2.5 GHz fractional-N frequency synthesizer for CDMA-2000." *Proceedings of the 27th European Solid-State Circuits Conference* (September), pp. 9–12.

Lee, M., Heidari, M., and Abidi, A. (2009). "A low-noise wideband digital phase-locked loop based on a course-fine time-to-digital converter with subpicosecond resolution." *IEEE Journal of Solid-State Circuits* 44(10) (October), 2808–2816.

Lindsey, W., and Chie C. (eds.) (1986). *Phase-Locked Loops.* New York: IEEE Press.

Liu, L., and Li, B. (2005). "Phase noise cancellation for a Σ-Δ fractional-N PLL employing a sample-and-hold element." *APMC 2005. Asia-Pacific Microwave Conference Proceedings* (available from IEEE Xplore), Vol. 5, December 4–7.

Lyons, R. (2004). *Understanding Digital Signal Processing*, 2nd ed. Upper Saddle River, NJ: Prentice Hall.

Mair, H., and Xiu, L. (2000). "An architecture of high-performance frequency and phase synthesis." *IEEE Journal of Solid-State Circuits* 35(6) (June), 835–846.

Margarit, M., Tham, J., Meyer, R., and Deen, M. (1999). "A low-noise, low-power VCO with automatic amplitude control for wireless applications." *IEEE Journal of Solid-State Circuits* 34(6), 761–771. In Ravazi [2003], p. 280.

Meninger, S., and Perrott, M. (2003). "A fractional-N frequency synthesizer architecture utilizing a mismatch compensated PFD/DAC structure for reduced quantization-induced phase noise." *IEEE Transactions on Circuits and Systems II: Analog and Digital Signal Processing* 50(11) (November), 839–849.

Miller, B., and Conley, R. (1991). "A multiple modulator fractional divider." *IEEE Transactions on Instrumentation and Measurements* 40(5) (June), 578–583.

Montres, G., and Parker, T. (1990). "Design techniques for achieving state-of-the-art oscillator performance." *Proceedings of the 44th Annual Symposium on Frequency Control,* pp. 522–535.

Pamarti, S., and Galton, I. (2003). "Phase-noise cancellation design tradeoffs in delta-sigma fractional-N PLLs." *IEEE Transactions on Circuits and Systems II: Analog and Digital Signal Processing* 50(11) (November), 829–838.

Pamarti, S., and Galton, I. (2007). "LSB dithering in MASH delta-sigma D/A converters." *IEEE Transactions on Circuits and Systems I: Regular Papers* 54(4) (April), 779–790.

Pamarti, S., Jansson, L., and Galton, I. (2004). "A wideband 2.4-GHz delta-sigma fractional-N PLL with 1-Mb/s in-loop modulation." *IEEE Journal of Solid-State Circuits* 39(1) (January), 49–62. See analysis in Pamarti and Galton [2003].

Park C.-H., Kim, O., and Kim, B. (2001). "A 1.8-GHz self-calibrated phase-locked loop with precise I/Q matching." *IEEE Journal of Solid-State Circuits* 36(5) (May), 777–783.

PLB: see Egan [1998] or Egan [2007].

Razavi, B. (ed.) (1996). *Monolithic Phase-Locked Loops and Clock Recovery Circuits, Theory and Design.* New York: IEEE Press.

Ravazi, B. (ed.) (2003). *Phase-Locking in High-Performance Systems.* Hoboken, NJ: Wiley.

Rhee, W., Song, B., and Akbar, A. (2000). "A 1.1-GHz CMOS fractional-N frequency synthesizer with a 3-b third-order ΣΔ modulator." *IEEE Journal of Solid-State Circuits* 35 (10) (October), 1453–1460.

Riley, T., and Kostamovaara, J. (2003). "A hybrid ΣΔ fractional-N frequency synthesizer." *IEEE Transactions on Circuits and Systems II: Analog and Digital Signal Processing* 50(4) (April), 835–844.

Riley, T., Copeland, M., and Kwasniewski, T. (1993). "Delta-sigma modulation in fractional-N frequency synthesis." *IEEE Journal of Solid-State Circuits* 28(5) (May), 553–559.

Rogers, J., Plett, C., and Dai, F. (2006). *Integrated Circuit Design for High-Speed Frequency Synthesis.* Boston, MA: Artech House.

Rohde, U., and Poddar, A. (2006). "Cost-effective VCOs replace power-hungry YIGs." *Microwaves and RF* (April), p. 84.

Rohde, U., and Poddar, A. (2008). "Miniature VCOs shrink wideband synthesizers." *Microwaves and RF* (December), p. 88.

Schreier, R., and Temes, G. (2005). *Understanding Delta-Sigma Data Converters.* Piscataway, NJ: Wiley.

Shu, K., and Sánchez-Sinencio, E. (2005). *CMOS PLL Synthesizers, Analysis and Design.* Springer.

Shu, K., Sanchez-Sinencio, E., Maloberti, F., and Eduri, U. (2001) "A comparative study of digital ΣΔ modulators for fractional-N synthesis." *The 8th IEEE International Conference on Electronics, Circuits and Systems*, Vol. 3, 1391–1394.

Smith, M. (1996). *Offset reference PLLs for fine resolution or fast hopping*, Technical Data Document AN1277, July, Phoenix, AZ: Motorola Semiconductor Products.

Song, J., and Park, I. (2010). "Spur-free MASH delta-sigma modulation." *IEEE Transactions on Circuits and Systems I: Regular Papers* 57(9) (September), 2426–2437.

Sotiriadis, P. (2006). "Diophantine frequency synthesis." *IEEE Transactions on Ultrasonics, Ferroelectrics, and Frequency Control* 53(11) (November), 1988–1998.

Sotiriadis, P. (2008a). "Cascaded diophantine frequency synthesis." *IEEE Transactions on Circuits and Systems I: Regular Papers* 55(3) (April), 741–751.

Sotiriadis, P. (2008b) "Diophantine frequency synthesis for fast-hopping, high-resolution frequency synthesizers." *IEEE Transactions on Circuits and Systems II: Express Briefs* 55(4) (April), 374–378.

Sotiriadis, P. (2008c). "Design and implementation of a forward two-PLL diophantine frequency synthesizer with 500× resolution improvement." *2008 IEEE International Frequency Control Symposium* (September), pp. 572–575.

Sotiriadis, P. (2010a). "Theory of flying-adder frequency synthesizers—part I: modeling, signals' periods and output average frequency." *IEEE Transactions on Circuits and Systems I: Regular Papers* 57(8) (August), 1935–1948.

Sotiriadis, P. (2010b). "Theory of flying-adder frequency synthesizers—part II: time- and frequency-domain properties of the output signal." *IEEE Transactions on Circuits and Systems I: Regular Papers* 57(8) (August), 1949–1963.

Staszewski, R. B., and Balsara, P. T. (2006). *All-Digital Frequency Synthesizer in Deep-Submicron CMOS*. Hoboken, NJ: Wiley.

Staszewski, R. B., Fernando, C., and Balsara, P. (2005). "Event-driven simulation and modeling of phase noise of an RF oscillator." *IEEE Transactions on Circuits and Systems I: Regular Papers* 52(4) (April), 723–733.

Staszewski, R. B., Waheed, K., Vermlapalli, S., Vallur, P., Entezari, M., and Eliezer, O. (2009). "Elimination of spurious noise due to time-to-digital converter." *IEEE Dallas Circuits and Systems Workshop Technical Program*, October 5.

Syllaios, I. L., Staszewski, R. B., and Balsara, P. T. (2008). "Time-domain modeling of an RF all-digital PLL." *IEEE Transactions on Circuits and Systems II: Express Briefs* 55(6) (June), 601–605.

Temporiti, E., Weltin-Wu, C., Baldi, D., Tonietto, R., and Svelto, F. (2009). "A 3 GHz fractional all-digital PLL with a 1.8 MHz bandwidth implementing spur reduction techniques." *IEEE Journal of Solid-State Circuits* 44(3) 824–834.

Tiebout, M. (2001). "Low-power low-phase-noise differentially tuned quadrature VCO design in standard CMOS." *IEEE Journal of Solid-State Circuits* 36(July), 1018–1024. In Ravazi [2003], p. 299.

Tokairin, T., Okada, M., Kitsunezuka, M., Maeda, T., and Fukaihi, M. (2010). "A 2.1-to-2.8 GHz all-digital frequency synthesizer with a time-windowed TDC." *IEEE International Solid-State Circuits Conference Digest of Technical Papers*, pp. 470–471.

Vamvakos, S. D., Staszewski, R. B., Sheba, M., and Waheed, K. (2006). "Noise analysis of time-to-digital converter in all-digital PLLs." *Proceedings of Fifth IEEE Dallas Circuits and Systems Workshop: Design, Application, Integration and Software (DCAS-06)* (October), p. 87–90.

Vancorenland, P., and Steyaert, M. (2002). "A 1.57-GHz fully integrated very low-phase-noise quadrature VCO." *IEEE Journal of Solid-State Circuits* 37(May), 653–656. In Ravazi [2003], p. 309.

van der Tang, J., van der Pepijn, D., and van Roermund, A. (2002). "Analysis and design of an optimally coupled 5-GHz quadrature *LC* oscillator." *IEEE Journal of Solid-State Circuits* 37(5) (May), 657–661. In Ravazi [2003], p. 305.

Weltin-Wu, C., Temporiti, E., Baldi, D., Cusmai, M., and Svelto, F. (2010). "A 3.5 GHz ADPLL with fractional spur suppression through TDC dithering and feedforward compensation." *IEEE International Solid-State Circuits Conference Digest of Technical Papers*, pp. 468–469.

Xiu, L. (2008). "The concept of time-average-frequency and mathematical analysis of flying-adder frequency synthesis architecture." *IEEE Circuits and Systems Magazine* Third Quarter, pp. 27–51.

Xiu, L., and You, Z. (2002). "A 'flying-adder' architecture of frequency and phase synthesis with scalability." *IEEE Transactions on Very Large Scale Integration (VLSI) Systems* 10(5) (October), 637–649.

Xu, L., Lindfors, S., Stadius, K., and Ryynänen, J. (2009). "A 2.4-GHz low-power all-digital phase-locked loop." *IEEE Custom Integrated Circuits Conference (CICC)*, pp. 331–334.

Yang, S., Chen, W., and Lu, T. (2010). "A 7.1 mW, 10 GHz all digital frequency synthesizer with dynamically reconfigured digital loop filter in 90 nm CMOS technology." *IEEE Journal of Solid-State Circuits* 45(3) (March), 578–586.

Ye, Z., and Kennedy, M. (2009a). "Hardware reduction in digital delta-sigma modulators via error masking—part I: MASH DDSM." *IEEE Transactions on Circuits and Systems I: Regular Papers* 56(4) (April), 714–726.

Ye, Z., and Kennedy, M. (2009b). "Hardware reduction in digital delta-sigma modulators via error masking—part II: SQ-DDSM." *IEEE Transactions on Circuits and Systems II: Express Briefs* 56(2) (February), 112–116.

Zannoth, M., Kolb, B., Fenk, J., and Weigel, R. (1998). "A fully integrated VCO at 2 GHz." *IEEE Journal of Solid-State Circuits* 33(12) (December), 1987–1991. In Ravazi [2003], p. 282.

Z-Communications, Inc. (2008). "Another major breakthrough in VCO technology." *Microwave Journal* (December), p. 122.

INDEX

Advanced Frequency Synthesis by Phase Lock, First Edition. William F. Egan.
© 2011 John Wiley & Sons, Inc. Published 2011 by John Wiley & Sons, Inc.